Lecture Notes in Computer Science

Lecture Notes in Artificial Intelligence 15465

Founding Editor

Jörg Siekmann

Series Editors

Randy Goebel, *University of Alberta, Edmonton, Canada*
Wolfgang Wahlster, *DFKI, Berlin, Germany*
Zhi-Hua Zhou, *Nanjing University, Nanjing, China*

The series Lecture Notes in Artificial Intelligence (LNAI) was established in 1988 as a topical subseries of LNCS devoted to artificial intelligence.

The series publishes state-of-the-art research results at a high level. As with the LNCS mother series, the mission of the series is to serve the international R & D community by providing an invaluable service, mainly focused on the publication of conference and workshop proceedings and postproceedings.

Lourdes Martínez-Villaseñor ·
Gilberto Ochoa-Ruiz · Martin Montes Rivera ·
María Lucía Barrón-Estrada ·
Héctor Gabriel Acosta-Mesa
Editors

Advances in Computational Intelligence

MICAI 2024 International Workshops

HIS 2024, WILE 2024, and CIAPP 2024
Tonantzintla, Mexico, October 21–25, 2024
Proceedings, Part II

 Springer

Editors
Lourdes Martínez-Villaseñor ⓘ
Universidad Panamericana
Mexico City, Distrito Federal, Mexico

Martin Montes Rivera ⓘ
Universidad Politécnica de Aguascalientes
Aguascalientes, Mexico

Héctor Gabriel Acosta-Mesa ⓘ
Universidad Veracruzana
Veracruz, Mexico

Gilberto Ochoa-Ruiz ⓘ
Instituto Tecnológico y de Estudios
Superiores de Monterrey
Zapopan, Mexico

María Lucía Barrón-Estrada ⓘ
TecNM-Instituto Tecnológico de Culiacán
Sinaloa, Mexico

ISSN 0302-9743   ISSN 1611-3349 (electronic)
Lecture Notes in Artificial Intelligence
ISBN 978-3-031-83881-1   ISBN 978-3-031-83882-8 (eBook)
https://doi.org/10.1007/978-3-031-83882-8

LNCS Sublibrary: SL7 – Artificial Intelligence

This Springer imprint is published by the registered company Springer Nature Switzerland AG
The registered company address is: Gewerbestrasse 11, 6330 Cham, Switzerland

If disposing of this product, please recycle the paper.

Preface

The Mexican International Conference on Artificial Intelligence (MICAI) is a yearly international conference series that has been organized by the Mexican Society for Artificial Intelligence (SMIA) since 2000. MICAI is a major international artificial intelligence (AI) forum and the main event in the academic life of the country's growing AI community.

This year, MICAI 2024 was organized by the Mexican Society for Artificial Intelligence (SMIA, Sociedad Mexicana de Inteligencia Artificial) in collaboration with the Instituto Nacional de Astrofísica, Óptica y Electrónica (INAOE) and the Universidad de las Américas Puebla (UDLAP).

The MICAI series website is www.MICAI.org. The website of the Mexican Society for Artificial Intelligence, SMIA, is www.SMIA.mx. Contact options and additional information can be found on these websites.

The conference, as is traditional, showcased a large variety of research fields and topics. Moreover, the conference included cutting-edge keynote lectures, as well as detailed paper presentations and comprehensive hands-on tutorials. Furthermore, thought-provoking panels and niche workshops provided a rich and exciting experience that aimed to cater to a wide audience.

Moreover, we continued the legacy of announcing the José Negrete Award, the SMIA Best Thesis in Artificial Intelligence Contest's results. This year, the historic and culturally rich city of Puebla was our chosen rendezvous.

MICAI conferences publish high-quality papers in all areas of AI and its applications. The proceedings of the previous MICAI events have been published by Springer in its Lecture Notes in Artificial Intelligence (LNAI) series (volumes: 1793, 2313, 2972, 3789, 4293, 4827, 5317, 5845, 6437, 6438, 7094, 7095, 7629, 7630, 8265, 8266, 8856, 8857, 9413, 9414, 10061, 10062, 10632, 10633, 11288, 11289, 11835, 12468, 12469, 13067, 13068, 13612, 13613, 14502, 14391, and 14392). Since its foundation in 2000, the conference has grown in popularity and improved in quality.

Three workshops were held jointly with the conference. The proceedings of the MICAI 2024 workshops are published in two volumes. The first volume contains 20 papers from the 17th Workshop of Hybrid Intelligent Systems (HIS 2024). The second volume contains 18 papers from the 17th Workshop on Intelligent Learning Environments (WILE 2024) and the 6th Workshop on New Trends in Computational Intelligence and Applications (CIAPP 2024).

These volumes will be of interest for researchers in all fields of artificial intelligence, students specializing in related topics, and the general public interested in recent developments in AI.

The MICAI workshops received for evaluation 58 submissions. From these submissions, 38 papers were selected for publication in these volumes after a double-blind peer-reviewing process and three reviews per submission. It was carried out by the Program Committee of the workshops. The acceptance rate was 65%.

HIS 2024

Hybrid Intelligent Systems (HIS) offer a compelling solution to complex challenges in various application domains, including biology, medicine, logistics, management, engineering, technology, social, and humanities. The 17th Workshop of Hybrid Intelligent Systems (HIS 2024) gathered relevant research associated with HIS and their capabilities for managing all these complex processes.

HIS 2024 was hosted by the Mexican Society of Artificial Intelligence (SMIA), nested within the Mexican International Conference on Artificial Intelligence (MICAI 2024). This year, the workshop featured 20 selected research articles of 32 received on the CMT platform. All the articles underwent a rigorous double-blind peer review with an acceptance rate of 62.5%, covering topics such as Machine Learning, Fuzzy Systems, Reasoning, Intelligent Control, Computer Vision, Optimization, and Expert Systems.

HIS 2024 presented the latest advancements in the field with two track sessions: the first featured articles with finished research, while the second consisted of posters that showcased proposals and prototypes for future research. The workshop's primary objective was to present expert model systems applied to specific research topics.

We sincerely thank Patricia Melin of the Tijuana Institute of Technology, our keynote speaker, for her valuable contribution to our event with the lecture "Hybrid Intelligent Models based on Neural Networks and Fuzzy Logic." HIS 2024 was enriched by her extensive experience and remarkable achievements in the field. We greatly appreciate her presence and the insight she shared with our audience.

We would also like to thank SMIA and the MICAI organizers for their continued collaboration in developing this prestigious event. It was our honor to contribute to the MICAI event with research in Hybrid Intelligent Systems.

WILE 2024

In these times when we witness the fantastic advance of science and technology around artificial intelligence, the multiplication of knowledge has become a challenge, not only for human beings but also for the most powerful computers. The assimilation of such knowledge requires the development of more effective learning methods for both humans and machines. These new methodologies, in order to be in line with the trend towards smarter systems, need to be intelligent enough to facilitate and adequately condition the learning process. This type of learning should not be limited to individual learning or to following current trends and practices, but should break with restrictive schemes and allow learning anywhere, in all kinds of situations and under any circumstances. This is why the first mental change must occur in the perception of classroom learning or solitary reading and one must think that learning must occur in the environment in which each person finds themselves.

This is why the Workshop on Intelligent Learning Environments (WILE) is organized to promote and encourage the development of methods and technologies in the living surroundings. This workshop presents new ideas and developments in all areas of academia, education, training, development, and research, using a wide variety of learning methods and artificial intelligence.

The aim of this workshop is to bring together active researchers and students in the field of Intelligent Learning Environments so that they can present and discuss innovative theoretical work and original applications, exchange ideas, establish collaboration links, discuss important recent achievements, and talk over the significance of results in the field to AI in general. Our goal for WILE 2024 was to give researchers a platform to showcase their work while also investigating new approaches to integrate AI techniques in the creation of educational systems.

We invited authors to submit papers presenting original and unpublished research in all areas related to:

- Web-based intelligent tutoring systems
- Intelligent learning management systems
- Affective tutoring systems
- Modeling, enactment, and intelligent use of emotion and affect
- Natural language and dialogue approaches to ILE design and construction
- Authoring tools in intelligent tutoring systems
- Learning companions
- Applications of cognitive science
- Semantic Web technologies
- Student modeling
- Sentiment analysis in educational applications
- Gamification and game-based learning
- Educational data mining
- Learning analytics

The entire submission, reviewing, and selection process, as well as preparation of the proceedings, was supported by Microsoft's Conference Management Toolkit. Many people took part in WILE 2024, and we are thankful for the cooperation of the Red-ICA (Conacyt Thematic Network in Applied Computational Intelligence) members who served on the Technical Committee, as well as members of The Mexican Society for Artificial Intelligence (SMIA Sociedad Mexicana de Inteligencia Artificial). As every year, SMIA 2024 was the appropriate host for this event.

CIAPP 2024

The 6th Workshop on New Trends in Computational Intelligence and Applications (CIAPP 2024) aimed to bring together researchers, students, and end users to explore the latest advances in computational intelligence. This year's workshop focused on innovative algorithms, applications in health sciences, and interdisciplinary approaches that use machine learning, computer vision, neural networks, and evolutionary computing. Participants had the opportunity to share their findings, engage in discussions, and start collaborations, fostering a vibrant community dedicated to pushing the boundaries of computational intelligence.

We thank the Sociedad Mexicana de Inteligencia Artificial (SMIA) for organizing the 23rd Mexican International Conference on Artificial Intelligence (MICAI 2024). Their dedication and hard work created a platform for researchers and students to share knowledge, foster collaboration, and advance the field of artificial intelligence.

We also extend our most sincere thanks and recognition to the Instituto Nacional de Astrofísica Óptica y Electrónica (INAOE) and the Universidad de las Américas Puebla (UDLAP) for their outstanding support and exceptional organization as a local committee of MICAI 2024. Their commitment to organizing this event created an attractive environment for collaboration and knowledge sharing. Thank you for your invaluable contributions to the success of CIAPP 2024.

We want to thank all the people involved in the organization of this conference: the authors of the papers published in these two volumes –it is their research work that gives value to the proceedings– and the organizers for their work. We thank the reviewers for their great effort spent on reviewing the submissions and the Program and Organizing Committee members.

A special acknowledgment to the local committee led by Jose Martinez-Carranza, whose meticulous coordination was instrumental in realizing MICAI 2024 in Tonantzintla, Puebla, Mexico. Our thanks extend to INAOE's director, David Sánchez de La Llave. We are also indebted to Luis Ernesto Derbez Bautista, Rector of UDLAP, for his invaluable assistance in securing the university facilities to complement the facilites of INAOE.

In addition, the success of this conference and the breadth of its program are a testament to the collaborative efforts of our organizers and committees, sponsors, and the invaluable support of the US Office of Naval Research Global (ONRG) with the grant award number N629092412100 from the agency GRANT14222939.

The entire submission, reviewing, and selection process, as well as preparation of the proceedings, was supported by Microsoft's Conference Management Toolkit (https://cmt3.research.microsoft.com/). Last but not least, we are grateful to Springer for their patience and help in the preparation of these volumes.

In conclusion, MICAI 2024 was more than just a conference. It was a confluence of minds, a testament to the indefatigable spirit of the AI community, and a beacon for the future of Artificial Intelligence. As you navigate through these proceedings, may you find inspiration, knowledge, and connections that propel you forward in your journey.

November 2023

Lourdes Martínez-Villaseñor
Gilberto Ochoa-Ruiz
Martin Montes Rivera
María Lucía Barrón Estrada
Efrén Mezura-Montes

Conference Organization

Conference Committee

General Chair

Lourdes Martínez-Villaseñor   Universidad Panamericana, Mexico

Program Chair

Gilberto Ochoa-Ruiz   Tecnológico de Monterrey, Mexico

Workshop Chair

Hiram Ponce   Universidad Panamericana, Mexico

Tutorials Chairs

Roberto Antonio Vázquez   Universidad La Salle, Mexico
Espinoza de los Monteros

Doctoral Consortium Chairs

Miguel González Mendoza   Tecnológico de Monterrey, Mexico
Juan Martínez Miranda   Centro de Investigación Científica y de Educación
Superior de Ensenada, Mexico

Keynote Talks Chairs

Gilberto Ochoa Ruiz   Tecnológico de Monterrey, Mexico
Iris Méndez   Universidad Autónoma de Ciudad Juárez, Mexico

Publication Chair

Hiram Ponce   Universidad Panamericana, Mexico

Financial Chairs

Hiram Calvo   Instituto Politécnico Nacional, Mexico
Lourdes Martínez-Villaseñor   Universidad Panamericana, Mexico

Grant Chair

Leobardo Morales   IBM, Mexico

Local Organizing Committee

Jose Martinez-Carranza (INAOE)
Jezabel Guzman-Zavaleta (UDLAP)
Caleb Rascón-Estebané (UNAM)
Gustavo Rodríguez-Gómez (INAOE)
Aldrich Alfredo Cabrera-Ponce (BUAP)
Brenda Cervantes-Cuahuey (INAOE)
Delia Irazú Hernández-Farias (INAOE)
Alejandro Gutiérrez-Giles (INAOE)
Leticia Oyuki Rojas-Perez (INAOE)

Program Committee

Alberto Ochoa-Zezzatti   Universidad Autónoma de Ciudad Juárez, Mexico
Aldo Marquez-Grajales   Instituto Tecnológico Superior de Xalapa, Mexico
Alexander Bozhenyuk   Southern Federal University, Russia
Andrés Espinal   Universidad de Guanajuato, Mexico
Angel Sánchez García   Universidad Veracruzana, Mexico
Anilu Franco   Universidad Autónoma del Estado de Hidalgo, Mexico
Antonieta Martinez   Universidad Panamericana, Mexico
Antonio Neme   UNAM, Mexico
Ari Barrera Animas   Universidad Panamericana, Mexico
Asdrúbal López Chau   Universidad Autónoma del Estado de México, Mexico
Belém Priego Sánchez   Universidad Autónoma Metropolitana Unidad Azcapotzalco, Mexico
Bella Martinez Seis   Instituto Politécnico Nacional, Mexico
Betania Hernandez-Ocaña   Universidad Juárez Autónoma de Tabasco, Mexico
Claudia Gómez   Instituto Tecnológico de Ciudad Madero, Mexico

Daniela Alejandra Ochoa	CentroGEO-CONACyT, Mexico
Dante Mújica-Vargas	CENIDET, Mexico
Diego Uribe	Tecnológico Nacional de México - ITL, Mexico
Eddy Sánchez-DelaCruz	Tecnológico Nacional de México - Campus Misantla, Mexico
Eduardo Valdez	Instituto Politécnico Nacional, Mexico
Efrén Mezura-Montes	Universidad Veracruzana, Mexico
Eloísa García-Canseco	Universidad Autónoma de Baja California, Mexico
Elva Lilia Reynoso Jardon	Universidad Autónoma de Ciudad Juárez, Mexico
Eric Tellez	CICESE-INFOTEC-CONACyT, Mexico
Ernesto Moya-Albor	Universidad Panamericana, Mexico
Félix Castro Espinoza	Universidad Autónoma del Estado de Hidalgo, Mexico
Fernando Gudino	UNAM, Mexico
Garibaldi Pineda Garcia	Applied AGI, UK
Genoveva Vargas-Solar	Grenoble Alpes University, CNRS, France
Gilberto Ochoa-Ruiz	Tecnológico de Monterrey, Mexico
Giner Alor-Hernandez	Tecnológico Nacional de México - ITO, Mexico
Guillermo Santamaría-Bonfil	BBVA México, Mexico
Gustavo Arroyo	Instituto Nacional de Electricidad y Energías Limpias, Mexico
Helena Gómez Adorno	IIMAS-UNAM, Mexico
Hiram Ponce	Universidad Panamericana, Mexico
Hiram Calvo	Instituto Politécnico Nacional, Mexico
Hugo Jair Escalante	INAOE, Mexico
Humberto Sossa	Instituto Politécnico Nacional, Mexico
Iris Iddaly Méndez-Gurrola	Universidad Autónoma de Ciudad Juárez, Mexico
Iskander Akhmetov	Institute of Information and Computational Technologies, Kazakhstan
Ismael Osuna-Galán	Universidad de Quintana Roo, Mexico
Israel Tabarez	Universidad Autónoma del Estado de México, Mexico
Jaime Cerda	Universidad Michoacana de San Nicolás de Hidalgo, Mexico
Jerusa Marchi	Federal University of Santa Catarina, Brazil
Joanna Alvarado Uribe	Tecnológico de Monterrey, Mexico
Jorge Perez Gonzalez	UNAM, Mexico
José Alanis	Universidad Tecnológica de Puebla, Mexico
José Martínez-Carranza	INAOE, Mexico
José Alberto Hernández-Aguilar	Universidad Autónoma del Estado de Morelos, Mexico
José Carlos Ortiz-Bayliss	Tecnológico de Monterrey, Mexico

Juan Villegas-Cortez	UAM - Azcapotzalco, Mexico
Juan Carlos Olivares Rojas	Tecnológico Nacional de México - ITM, Mexico
Karina Perez-Daniel	Universidad Panamericana, Mexico
Karina Figueroa Mora	Universidad Michoacana de San Nicolás de Hidalgo, Mexico
Leticia Flores Pulido	Universidad Autónoma de Tlaxcala, Mexico
Lourdes Martínez-Villaseñor	Universidad Panamericana, Mexico
Luis Torres-Treviño	Universidad Autónoma de Nuevo León, Mexico
Luis Luevano	Institut National de Recherche en Informatique et en Automatique, France
Mansoor Ali Teevno	Tecnológico de Monterrey, Mexico
Masaki Murata	Tottori University, Japan
Miguel Gonzalez-Mendoza	Tecnológico de Monterrey, Mexico
Miguel Mora-Gonzalez	Universidad de Guadalajara, Mexico
Mukesh Prasad	University of Technology Sydney, Australia
Omar López-Ortega	Universidad Autónoma del Estado de Hidalgo, Mexico
Rafael Guzman-Cabrera	Universidad de Guanajuato, Mexico
Rafael Batres	Tecnológico de Monterrey, Mexico
Ramon Brena	Instituto Tecnológico de Sonora, Mexico
Ramón Zatarain Cabada	Tec Culiacán, Mexico
Ramón Iván Barraza-Castillo	Universidad Autónoma de Ciudad Juárez, Mexico
Roberto Antonio Vasquez	Universidad La Salle, Mexico
Rocio Ochoa-Montiel	Universidad Autónoma de Tlaxcala, Mexico
Ruben Carino-Escobar	Instituto Nacional de Rehabilitación - Luis Guillermo Ibarra Ibarra, Mexico
Sabino Miranda	INFOTEC-CONACyT, Mexico
Saturnino Job Morales	Universidad Autónoma del Estado de México, Mexico
Segun Aroyehun	University of Konstanz, Germany
Sofía Galicia Haro	Sistema Nacional de Investigadoras e Investigadores, Mexico
Tania Ramirez-delReal	CentroGEO-CONACyT, Mexico
Vadim Borisov	Branch of National Research University "Moscow Power Engineering Institute" in Smolensk, Russia
Valery Solovyev	Kazan Federal University, Russia
Vicenc Puig	Universitat Politècnica de Catalunya, Spain
Vicente Garcia Jimenez	Universidad Autónoma de Ciudad Juárez, Mexico
Victor Lomas-Barrie	IIMAS-UNAM, Mexico

Workshops Organization

HIS 2024 Organizing Committee

General Chairs

Martín Montes Rivera	Universidad Politécnica de Aguascalientes, Mexico
Carlos Alberto Ochoa Zezzatti	Universidad Autónoma de Ciudad Juárez, Mexico
Daniela Paola López Betancur	Universidad Autónoma de Zacatecas, Mexico
Carlos Alejandro Guerrero Méndez	Universidad Autónoma de Zacatecas, Mexico
José Alberto Hernández Aguilar	Universidad Autónoma del Estado de Morelos, Mexico

Program Committee

Edgar Gonzalo Cossio Franco	Instituto de Información Estadística y Geográfica de Jalisco, Mexico
Humberto Velasco Arellano	Universidad Politécnica de Aguascalientes, Mexico
Daniela Paola López Betancourt	Universidad Politécnica de Aguascalientes, Mexico
Carlos Alejandro Guerrero Méndez	Universidad Autónoma de Zacatecas, Mexico
Humberto Muñoz Bautista	Universidad Tecnológica Metropolitana de Aguascalientes, Mexico
Miguel Ángel Ortiz Esparza	Universidad Autónoma de Aguascalientes, Mexico
Himer Avila-George	Universidad de Guadalajara, Mexico
Alejandro Padilla Díaz	Universidad Autónoma de Aguascalientes, Mexico
Carlos Alberto Lara Alvarez	CIMAT Zacatecas, Mexico
Roberto Antonio Contreras Masse	Instituto Tecnológico Ciudad Juárez, Mexico
Irma Yazmín Hernández Báez	Universidad Politécnica el Estado de Morelos, Mexico

WILE 2024 Organizing Committee

General Chairs

María Lucía Barrón Estrada	TecNM-Instituto Tecnológico de Culiacán, Mexico
Ramón Zatarain Cabada	TecNM-Instituto Tecnológico de Culiacán, Mexico
Yasmín Hernández Pérez	TecNM-Cenidet, Mexico
Carlos A. Reyes García	INAOE, Mexico
Karina Mariela Figueroa Mora	Universidad Michoacana de San Nicolás de Hidalgo, Mexico

Program Committee

Ramón Zatarain Cabada	Instituto Tecnológico de Culiacán, Mexico
María Lucía Barrón Estrada	Instituto Tecnológico de Culiacán, Mexico
Yasmín Hernández Pérez	Cenidet, Mexico
Karina Mariela Figueroa Mora	UMSNH, Mexico
Carlos A. Reyes García	Instituto Nacional de Astrofísica, Óptica y Electrónica, Mexico
Giner Alor Hernández	Instituto Tecnológico de Orizaba, Mexico
Miguel Pérez Ramírez	Instituto Nacional de Electricidad y Energías Limpias, Mexico
Jaime Muñoz Arteaga	Universidad Autónoma de Aguascalientes, Mexico
Rafael Morales Gamboa	Universidad de Guadalajara, Mexico
Guillermo Santamaría Bonfil	BBVA, Mexico
Carlos Alberto Lara Álvarez	CIMAT Zacatecas, Mexico
Hugo Arnoldo Mitre Hernández	CIMAT Zacatecas, Mexico
María Elena Chávez Echeagaray	Arizona State University, USA
María Blanca Ibáñez Espiga	Universidad Carlos III de Madrid, Spain
Alicia Martínez Rebollar	Cenidet, Mexico
María Lucila Morales Rodríguez	Instituto Tecnológico de Cd. Madero, Mexico
Héctor Rodríguez Rangel	Instituto Tecnológico de Culiacán, Mexico
Julieta Noguez Monroy	Tecnológico de Monterrey, Mexico
Samuel González López	Instituto Tecnológico de Nogales, Mexico
Raúl Oramas Bustillos	Universidad Autónoma de Occidente, Mexico
José Mario Ríos Félix	Instituto Tecnológico de Culiacán, Mexico
Luis Alberto Morales Rosales	Universidad Michoacana de San Nicolás de Hidalgo, Mexico
Maritza Bustos López	Instituto Tecnológico de Orizaba, Mexico

CIAPP 2024 Organizing Committee

General Chairs

Héctor Gabriel Acosta Mesa   Universidad Veracruzana, Mexico
Marcela Quiroz Castellanos   Universidad Veracruzana, Mexico
Rocío Erandi Barrientos Martínez   Universidad Veracruzana, Mexico
Efrén Mezura Montes   Universidad Veracruzana, Mexico

Program Committee

Martha Lorena Avendaño   Universidad Veracruzana, Mexico
Aldo Márquez Grajales   Universidad Veracruzana, Mexico
Nancy Pérez Castro   Universidad de Papaloapan, Mexico
Rafael Rivera-López   Tecnológico Nacional de México-Instituto
                                            Tecnológico de Veracruz, Mexico
Guillermo Hoyos-Rivera   Universidad Veracruzana, Mexico
Adriana L. López Lobato   Universidad Veracruzana, Mexico
Octavio Ramos Figueroa   Universidad de Xalapa, Mexico
Jesús Adolfo Mejía de Dios   Universidad Autónoma de Coahuila, Mexico
Mario Graff Guerrero   INFOTEC Aguascalientes, Mexico
José Luis Morales Reyes   Universidad de Xalapa, Mexico

Contents – Part II

CIAPP 2024

Contents – Part I

WILE 2024

Expert System for Teaching Classification Systems Workflow

Christian Sánchez-Sánchez(✉)

Departamento de Tecnologías de la Información, Universidad Autónoma
Metropolitana Unidad Cuajimalpa, Mexico City, Mexico
csanchez@cua.uam.mx

Abstract. Artificial intelligence (AI) is increasingly present in our lives,
and it is generating a higher demand for human resources with knowledge
about this area of computing. It is essential to have tools or resources that
help in learning AI. This article presents an expert system that supports
teaching the immersed process (workflow) in developing systems that
use machine learning, particularly classification. This system is based
on rules and preconditions guiding the development process, validating
that the functions are executed logically and indicating errors made. The
system architecture allows users to add or remove functions and precondition rules. One key role of the system is to guide users in identifying
and solving errors. The qualitative evaluation showed that the students,
after using the system for a few weeks, were able to confidently identify
and solve errors when implementing practices that involved this type of
development process.

Keywords: Teaching Systems · Expert System for Teaching ·
Classification Workflow

1 Introduction

It is increasingly common for Artificial Intelligence (AI) to be present in our
lives. Machine Learning(ML) is one of the branches of artificial intelligence that
has grown the most and has allowed for an increase in the generation of available
applications. According to Samuel [1], Machine Learning is "The field of study
that offers computers the ability to learn without necessarily being explicitly
programmed".

A study published in 2018 by Indeed [2], a job search engine, shows how
in recent years jobs related to Artificial Intelligence have been increasing, such
as those that demand titles of machine learning engineer, data scientist, and
computational linguist. A recent O'Reilly survey revealed that nearly 50% of
enterprises view skill shortages as a significant obstacle to AI adoption. Early IA
learning exposure not only prepares individuals for the growing job market but
also equips them with the tools to understand and contribute to the technology
that shapes our future [3].

L. Martínez-Villaseñor et al. (Eds.): MICAI 2024, LNAI 15465, pp. 3–14, 2025.
https://doi.org/10.1007/978-3-031-83882-8_1

Different topics, such as concepts, definitions, and languages, can be found in ML teaching, and each of them implies a different degree of complexity. During the Machine Learning course teaching, when students begin to program applications that use this type of AI, they commonly also present problems in developing their programs. For example, some of the problems they present are: they do not apply the same processing to the test set as the training set, they do not know when to apply standardization or normalization, when to select attributes, and what data the prediction should be made to evaluate one model among others later. That is why workflow knowledge is essential for creating systems using Machine Learning.

This article presents an expert system that supports the teaching of Workflow to create this type of software. The system is divided into two forms of operation, configured by the teacher. The first "autopilot mode", which is the first way in which students use the system, as it guides the process of creating and evaluating an ML model, and another "free run mode" visualizes all the functionalities of the system and allows the student to freely selects the functions they want to apply in the Workflow; throwing error codes when students make mistakes so that they know where to correct to reach the goal of creating and evaluating an ML model.

The following article is structured as follows: Sect. 2 shows the related works; Sect. 3 shows information about the Python API, pros and cons, and why it is essential to know about the development workflow. Section 4 describes the phases of the workflow for the development of an ML system; the functions of the basic version of the system are described in Sect. 5. In Sect. 6 some rules that allow the system to guide or evaluate during the workflow are shown. Sections 7 give details of the system design, while Sect. 8 comments on what was observed when used by students. The last section talks about the conclusions and future work.

2   Related Work

Teaching fundamental AI (including Machine Learning) concepts and techniques has traditionally been done only in higher education [4]. Many of the universities that teach courses related to AI offer formal courses on ML; an example of this is the online course taught by Harvard [5] or MIT [6]. On the other hand, there are web pages and applications that, in addition to allowing self-taught learning, help to test and reinforce what has been learned.

One of the most famous websites for teaching Machine Learning is Kaggle [7], a platform for data scientists and machine learning practitioners. Kaggle offers courses and allows users to remotely obtain or publish datasets, models, and notebooks using GPUs. It requires knowledge of Python programming and certain mathematical concepts.

Google [8] also offers a course organized into 12 modules with more than 100 exercises, real-world examples, and explanatory videos. This page allows users to have interactive visualizations of many of the concepts and algorithms they teach.

Speaking of more specific applications, Apps for Good: Machine Learning in a Day [9] is a Website that offers an overview of the application and its impact on the life of machine learning, reviewing case studies and exploring ethical issues.

There is also MIT App Inventor [10], a course that introduces the fundamentals of machine learning and guides students in developing their applications, but it focuses only on image classification.

The eCraft2Learn project [11] created extensions for the Snap! Programming language, allowing children and non-expert programmers to develop small AI programs.

Concerning the design, training, and evaluation of neural networks, several pages allow it to be done graphically and on GPUs remotely, such as A Neural Network Playground [12]. However, it is important to mention that knowledge of the subject is necessary.

There are also proposals related to learning AI in specific applications, such as the work presented by Narahara and Kobayashi [13], which uses a framework of interactive educational modules to introduce concepts in AI and robotics using an autonomous toy car equipped with a camera controlled by a Raspberry Pi.

Although several platforms support the teaching of ML, such as those shown above, one that supports understanding the workflow of creating systems based on Machine Learning is also needed, and this article presents a proposal.

3   The Scikit-Learn Python API

Several functions or methods are used in developing a Machine Learning system, many of them following the design of an API. This is the case of the functions offered by the Scikit-learn Python library; other similar libraries have adopted the same design.

The Scikit-Learn (sklearn) API is intuitive and consistent, with three main object types:

Estimators: Objects that estimate parameters in a dataset using the fit() method, which can take the dataset and its labels.

Transformers: Estimators that can also transform the dataset with the transform() method. They include the fit_transform() method to optimize the process.

Predictors: Estimators that make predictions with the predict() method and evaluate the quality of the predictions with score().

Additionally, to connect this type of objects, Scikit-Learn has **Pipelines**. A Pipeline allows for the sequential application of various transformers. Intermediate steps must implement fit and transform methods, whereas the final estimator only needs to implement fit. The logical sequence of what is executed is the application programmer's responsibility, and this also requires knowledge of the processes or workflow.

4   Workflow for Developing an ML System

According to IBM [14], "A workflow is a system for managing repetitive processes and tasks that occur in a particular order".

Developing ML applications is an iterative process involving some steps(some of them in sequential order), including [15]: Requirements Analysis, Data Management, Feature Engineering, Model Learning, Model Evaluation, and Model Deployment. The extracted workflow phases are described next:

- **Data Collection:** This involves gathering and analyzing data from various sources to solve research problems, answer questions, assess outcomes, and predict trends and probabilities.
- **Data Management:** The data is validated and cleaned and may undergo preprocessing to transform the raw data. The specific type of data preparation depends on the machine-learning task
- **Exploratory Data Analysis (EDA):** It guides the optimal manipulation of data sources, facilitating the discovery of patterns, identification of anomalies, hypothesis testing, and validation of assumptions.
- **Model Creation (Model Learning):** In this phase, the model is selected, and the hyperparameters are configured so that the training data is later used to pass it to the selected algorithm and create the model.
- **Model Evaluation:** It employs various evaluation metrics to assess a machine learning model's performance, highlighting its strengths and weaknesses.
- **Model Deployment:** Making the model available in a production environment.

Now that the phases have been described, the following section declares some of the functions or methods contained in the system. It is essential to mention that in the first version of the system, only the functions to generate and evaluate Classification models are included.

5   Available Functions for the Basic Version of the System

In the Data Collection stage, students must obtain the data they will work with. Initially, they are provided with specific datasets with which they can begin to experiment.

The tables of each phase (Tables 1, 2, 3 and 4) show the system function's name and a brief description of it.

Functions related to Data Management. In the first version of the system for this phase, there are 16 functions mainly to preprocess the data, part of them are listed in Table 1.

EDA Related Functions. In this phase, 14 methods that show statistical data such as mean, standard deviation, quartiles, number of records, and column data types can be observed; some listed in Table 2. This type of analysis can

Table 1. Data Management functions for the basic system

Name	Description
read_csv	Allows the user to upload a comma separated values file
get_KBestChiSqrd	Selects, using chi-square, the best "k" characteristics
minmax_scaler	Rescales all values between zero and one
train_test_split	Gets training and testing data from the predictors and classes

be performed before, after, or during Data Management. In general, built-in functions can extract information from particular objects but do not modify their state.

Table 2. EDA functions for the basic system

Name	Description
describe	Shows basic statistics of numerical characteristics (columns)
value_counts	Returns the count of the values of the characteristic or column
get_correlated_attrs	Gets the pairs of correlated attributes over a given threshold
check_for_duplicates	Check for duplicate records, instances, or items

Some of the methods mentioned in this section are related to functions commonly used in Python and use parameters to execute e.g. get_KBestChiSqrd requires "k_selected" the number of attributes to select. The system allows the user to give values to some of the parameters of the methods they wish to use.

Functions for Model Learning. In the basic version (see Table 3), a single algorithm (Decision Tree) is presented that obtains a model from the training data, but the objective is for the student to at least add more and test them.

Table 3. Model Learning functions for the basic system

Name	Description
DT_fit	Obtains a classification model using the Decision Tree(DT)algorithm

Functions for Model Evaluation. Those are 4 functions (some of them in Table 4) that allow for predicting the test set and subsequent application of some evaluation metrics. In the first version, only metrics for evaluating classification, such as precision, recall, and f1, are included.

It is worth mentioning that in the first version of the system, the integrated functions for this phase only reached the evaluation of the model using the precision, recall, and f1 metrics, letting the students implement the last Model Deployment phase.

Table 4. Model Evaluation functions for the basic system

Name	Description
DT_predict	Predicts the class of the testset instances using the DT model
DT_predict_proba	Predicts the probability of the testset using the DT model
precision_recall_score	Prints the precision and recall of the testset

6   Precondition Rules

It is essential to mention that in the workflow, not necessarily all available functions must be run; some do not have a run order, and some have to be executed in a certain order to be able to change phases. Rules verify certain preconditions related to states of objects or previously executed functions to verify that functions are executed correctly. These rules were obtained from the experience of previous courses on how students created this type of system and their main errors.

Below are some of the basic rules that check some states of global variables in order to let some functions be executed.

```
if csv_file_name is not None: EXEC("read_csv")
if dataset is not None: EXEC("train_test_split")
if X_training is not None and y_training is not None: EXEC("
    DT_fit")
if X_test is not None and y_test is not None: EXEC("DT_predict")
```

EXEC allows the function to be executed if and only if the condition is met. For example, DT_fit, the function that trains a decision tree, is only executed if it has the input predictors (X_training) and its classes (y_training).

As previously mentioned, the system has two types of execution: the first (autopilot mode) guides the student during the workflow, enabling and disabling functions so that the student can complete the workflow following a logical sequence. For doing this, precondition rules were defined; some of them are listed next:

```
if RUN("train_test_split"): SET_ENABLE("normalizer", True)
if RUN("train_test_split"): SET_ENABLE("DT_fit", True)
if RUN("DT_fit"): SET_ENABLE("DT_predict", True)
if RUN("DT_predict"): SET_ENABLE("f1_score", True)
```

In the previous rules, SET_ENABLE is a function that allows to enable (True) or disable(False) the display of functions in the menu, while RUN verifies whether a function was executed. For example, the last two rules, related to classification validation (precision_recall_score or f1_score), are only activated if DT_predict predictions were previously made on the test set.

These rules allow backward chaining and features can only be enabled if a rule was previously met. For example, in order for the function to make predictions(DT_predict) for the test set to be enabled, DT_fit must have previously

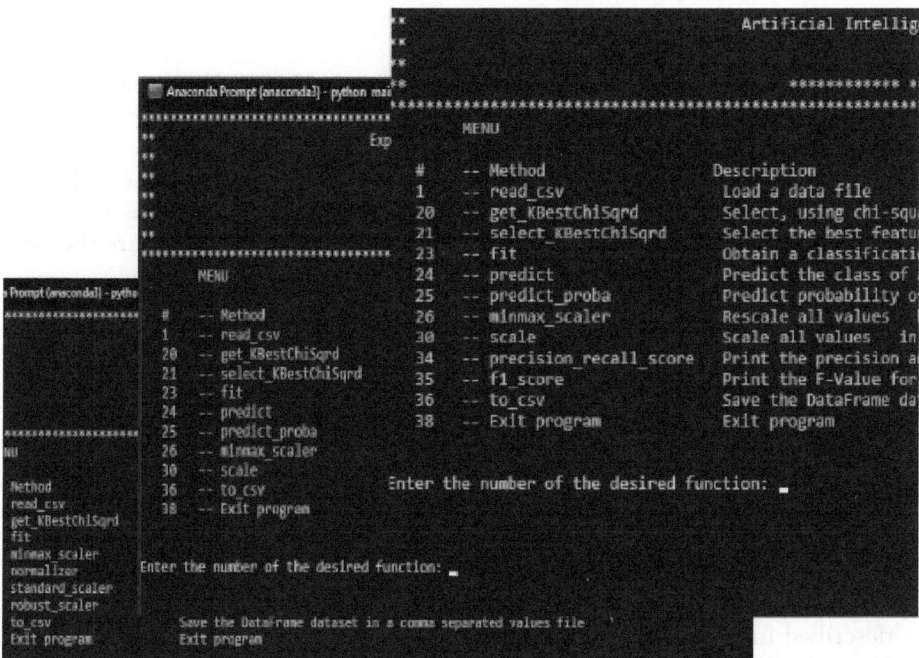

Fig. 1. Three sub-menus generated during the "autopilot mode" workflow

been executed, and for DT_fit to have been enabled, train_test_split and so on must have been previously executed. In the case of "autopilot mode," according to these rules, some functions can change state or substate in the workflow; when executed, they cause the system menu to also change, including other functions of the following state, as seen in the Fig. 1. Doing backward chaining is also how it is verified that there has been a logical sequence when the system is run in "free run mode".

The other functions are enabled if and only if read_csv was run, for example, and their rules are similar to the next example

```
if RUN("read_csv"): SET_ENABLE("corr", True)
```

Some rules turn off the options in the menu when some methods are executed; for example

```
if RUN("drop_na"): SET_ENABLE("drop_na", False)
```

In the previous example, once the records with null or NA values were deleted, it is not necessary to have the function active since once they are deleted, these values will no longer exist unless more data is loaded.

Figure 1 shows an example of a sequence of three screens with different sub-menus generated as the student uses functions and according to the rules so that the student can follow the workflow when using the "autopilot mode".

The following section briefly describes the system architecture.

7   System Architecture

The system operates under a console and must be installed on the student's computer. The system architecture is made up of three main components: the System Modules, which are the core of the system and allow the loading of functions dynamically according to the information in the database (DB); the DB, which contains the information of the methods and functions as well as the precondition rules and error codes; and the system functions, which are the basic functions plus those added by users so that they appear in the system.

7.1   System Modules

The System Modules contain the main classes for the execution of the system, and these are:

- **Main:** The Main Program is responsible for creating the user menu and invoking all the classes described below and the System Functions.
- **Loader:** Class loader uses reflection to load other classes and their methods or functions dynamically.
- **Requirement:** Requirements evaluator verifies that the condition rules described in Sect. 7 are met.
- **Functions Aggregator:** Functions Aggregator allows functions to be added to the system, and this module allows the incorporation of new functionalities by students.
- **DBManager:** Manages the communication with a Database that stores information about the classes and functions of the system. The architecture of the DB is described in the next section.
- **Parameter Config:** The Parameter Configurator allows parameters to be captured to execute the functions.

7.2   Global Variables to Avoid Message Passing

These variables are the parameters that are common in several functions, in most phases. In Python, they are the main objects used in the process of developing software using scikit-learn and most of the time, similar functions, to those of the expert system also use these objects. Table 5 shows some of the names of the variables and a brief description.

Table 5. Some Global Variables

Variable Name	Description
X_training	Training Predictors
y_training	Training Labels
X_test	Test Predictors
y_test	Test Labels

```
19    -- drop_duplicates        Remove duplicate data
20    -- get_KBestChiSqrd        Select, using chi-square, the best "k" characteristics
21    -- select_KBestChiSqrd     Select the best features, according to chi-square, from the "test data"
22    -- set_current_selector    Set a feature selector
23    -- fit                     Obtain a classification model using the DT algorithm
24    -- predict                 Predict the class of a "test set" using the obtained DT model
25    -- predict_proba           Predict probability of the "test set" of belonging to each class using DT
26    -- minmax_scaler           Rescale all values   between zero and one
27    -- normalizer              Normalize all values   to the unit norm
```

Fig. 2. An excerpt of the free run mode menu

```
Run Methods:
        1.-read_csv
        15.-get_class_and_data
        33.-train_test_split
        20.-get_KBestChiSqrd
        21.-select_KBestChiSqrd
        27.-normalizer
        30.-scale
        23.-fit
        24.-predict
        35.-f1_score
```

Fig. 3. Exiting the program and showing the run methods

7.3   System Functions

This module contains all system functions for the workflow, for example, those described in Sect. 6 and others added by the user.

Figure 2 shows an excerpt of the "free run mode" system screen, which contains all the functions available in the basic version. The system is run under the console by entering the selected function number and pressing "Enter", where function 38 exits the program.

It is essential to mention that whenever the program is exited, the console shows the sequence in which the methods were executed (See Fig. 3), and also, for each use, the system saves a log where the sequence of methods is stored; allowing the teacher to evaluate or detect errors committed subsequently.

7.4   Database Design

The Database is composed of 4 tables: codes, methods, methodrequirement and requirements.

The table called "codes" has the error codes (some in Table 7). These are described by an identifier and a textual description.

The table called "methods" stores the methods added to the system. Each method has a unique identifier, the name of the module or package in which it is located, the name of the function, the parameters it occupies, a description in natural language of what the function does, and the name of the function as we want it to appear in the system menu.

The "methodrequirement" table stores the precondition rules (defined in Sect. 6) that must have been executed so that the added function can be executed.

Finally, the table "requirement" contains the messages to be displayed if any precondition methods are not met (See Table 6).

When the system is used in "free run mode" and the student makes an error trying to execute functions where one of the precondition rules, shown in Sect. 6, is not met, a precondition not met message is launched in the system (see some in Table 6) which indicates the error or suggests which method should have been executed to comply with the said rule.

Table 6. Some Precondition not met messages

ID	Message	Required method
23	You should have obtained a classification model	fit()
24	The classes of the training set should have been predicted	predict()
26	The data should have been rescaled	minmax_scaler()
33	You should have separated the data into training and testing	train_test_split()

Error codes (some of them in Table 7) may also appear when the code contains an exception or when the parameters of the functions or global variables fail.

Table 7. Error Codes

Error Code	Description
1	Failure by exception
204	The class and predictors of the data have not been separated
300	The model has not been trained
302	The parameters necessary to predict were not passed

8   The Usage of the System by Students

Unfortunately, not much testing of the system has been done. The system was used to teach a Machine Learning course and advise students, totaling 14 students. The system began to be used after the topic of decision trees was discussed in the first third of the course.

Students began using the system in "autopilot mode" and were offered some example datasets like iris and weather. The objective was to use the software repeatedly to identify functions that changed the system menus(or submenus) and identify workflow phases. The practices aimed to use the system from loading the data to get a classification model and evaluate the model. While using the system, they were asked different requirements such as selecting attributes, other times normalizing or standardizing, and so on. So that they could make various combinations of functions. The student was graded by reviewing the sequence of executed functions that appeared on the console when they exited the system.

When they used the other mode, "free run mode", they were also asked to generate a classification model evaluated with a series of requirements. They were asked to write down the errors made in the process and submit the list of mistakes and how they corrected them at the end of the practice.

Regrettably, a quantitative evaluation of the software usage could not be done. Still, after a couple of weeks of using this system, compared with previous courses, an improvement was observed in developing the practices generated by programming in Python; it took students less time to develop the practices (identifying functions and workflow phases), and they made fewer common mistakes like not applying the same processing to the test set as the training set, not evaluating on a test set and so on. Even at this stage, the teacher was not consulted about how to solve problems related to the workflow.

Students also showed their commitment to learning by identifying errors that could have been generated by not meeting a precondition rule, like those listed in Sect. 6. This task, which required a little more time, demonstrated their willingness to invest extra effort in understanding the system and its requirements.

In the end, they were asked about the advantages they saw in the system. They most agreed that "it is not necessary to program while learning from the workflow". Their main criticism was that "the system works through a console", which they don't find very attractive.

9   Conclusions and Future Work

This article presents an expert system that supports teaching the immersed process (workflow) in developing systems that use machine learning, specifically classification. This system is based on precondition rules that guide the development process. Promising results were found since, after a couple of weeks of using this system, students made fewer errors related to the workflow. However, more tests need to be done with other students, and quantitative studies must be designed.

Implementing a more attractive GUI for the system is proposed. In the first version, because the system is standalone, its computing power depends on the characteristics of the computer where it runs in the course, and only simple applications were made. Nevertheless, online functions that run on a server with better characteristics (processing and GPU) for the system will be developed. This can help students make systems with more extensive data sets that they can access outside school.

References

1. Samuel, A.L.: Some studies in machine learning using the game of checkers. IBM J. Res. Dev. **3**(3), 210–229 (1959)
2. Culbertson, D.: Demand for AI talent on the rise - indeed hiring lab. https://www.hiringlab.org/2018/03/01/demand-ai-talent-rise/. Accessed 16 Sept 2024
3. Intel AI: AI goes to high school. Forbes. https://www.forbes.com/sites/insights-intelai/2019/05/22/ai-goes-to-high-school/#68826e3f1d0c. Accessed 16 Sept 2024
4. Torrey, L.: Teaching problem-solving in algorithms and AI. In: Proceedings of the AAAI Conference on Artificial Intelligence, vol. 26, no. 3, pp. 2363–2367 (2012)
5. Harvard: Machine Learning and AI with Python. https://pll.harvard.edu/course/machine-learning-and-ai-python. Accessed 16 Sept 2024
6. MIT: MITx: Machine Learning with Python: from Linear Models to Deep Learning. https://www.edx.org/learn/machine-learning/massachusetts-institute-of-technology-machine-learning-with-python-from-linear-models-to-deep-learning. Accessed 16 Sept 2024
7. Kaggle: Kaggle. https://www.kaggle.com/. Accessed 16 Sept 2024
8. Google: Machine Learning. https://developers.google.com/machine-learning/crash-course?hl=es-419. Accessed 16 Sept 2024
9. Apps for Good: Machine learning in a day. https://www.appsforgood.org/courses/ml-in-a-day. Accessed 16 Sept 2024
10. MIT: Introduction to Machine Learning: image classification. https://appinventor.mit.edu/explore/resources/ai/image-classification-look-extension. Accessed 16 Sept 2024
11. eCraft2Learn: Enabling children and beginning programmers to build AI programs. https://ecraft2learn.github.io/ai/. Accessed 16 Sept 2024
12. Carter, S., Smilkov D.: Tensorflow - Neural Network playground. https://playground.tensorflow.org/. Accessed 16 Sept 2024
13. Narahara, T., Kobayashi, Y.: Personalizing homemade bots with plug-and-play AI for STEAM education. In: SIGGRAPH Asia 2018 Technical Briefs, pp. 1–4 (2018)
14. IBM: What is a workflow?. https://www.ibm.com/topics/workflow. Accessed 16 Sept 2024
15. AWS: Building a machine learning application - Amazon Machine Learning. https://docs.aws.amazon.com/machine-learning/latest/dg/building-machine-learning.html. Accessed 16 Sept 2024

Enhancing Student Theses with Advanced Text Analysis Using NLP and Pre-trained Models

Maximiliano Ponce Marquez[1], Samuel González-López[1(✉)],
Jesús Raúl Cruz Rentería[1], Gilberto Borrego Soto[2],
Manuel Omar Meranza Castillon[1], and Guillermina Muñoz Zamora[1]

[1] Tecnológico Nacional de México Campus Nogales, Nogales, Sonora, Mexico
`samuel.gl@nogales.tecnm.mx`
[2] Instituto Tecnológico de Sonora, Obregon, Sonora, Mexico

Abstract. Writing a thesis can be challenging for university students, mainly due to the need for specific language tools in Spanish. This study presents an online platform that assists the thesis writing process by analyzing lexical richness, using Variety, Density, and Sophistication. Additionally, the platform includes a section to identify the structure of objectives and answer critical questions about what is going to be done? (Q1), What will it be done for? (Q2) And how will this action be carried out? (Q3). This platform aims to improve the quality of university theses. During the development, a usability survey was conducted with students, achieving an average rating of 8.12 out of 10, which allowed for identifying and correcting details for improvement. One of our work's contributions is using current web technologies combined with pre-trained BERT models, which will enable quick and easy user analysis.

Keywords: BERT models · spanish theses · lexical richness · lexical variety · lexical density · lexical sophistication · objective analysis · NLP

1 Introduction

Currently, in the midst of the digital age where information and technology play essential roles in our daily lives, the process of writing an academic thesis remains a significant challenge for university students. Although there are numerous tools available to analyze document writing in an academic context, most of them are designed in English [1,2].

Some of these tools include grammar checkers, which are used to analyze and correct grammatical errors, they are commonly found in web browsers and by default in text editors. There are significant advances in research and applications in Deep Learning, these tools have become more powerful. In [3], it is highlighted how the incorporation of Microsoft's grammatical verification in MS Word in 2004 seemed to hinder innovation in this field. However, twelve years later, in 2016, it is noted how new competitors, such as Grammarly, managed to establish themselves in the market.

L. Martínez-Villaseñor et al. (Eds.): MICAI 2024 Workshops, LNAI 15465, pp. 15–24, 2025.
https://doi.org/10.1007/978-3-031-83882-8_2

In addition to these types of tools, there are also some designed to enhance the creativity of writers using artificial intelligence, such as the work done in [4], where "Wordcraft", a text editor with integrated AI-powered writing assistance tools, was utilized. There is potential in this technology to make parts of the creative writing process easier, faster, and more enjoyable for both experienced writers and amateurs. To realize this potential, developers of AI writing tools must focus on the specific needs and desires of writers.

This work is a continuation of the platform created in [5], specifically focused on thesis writing. To improve and assist students, a web platform called TURET 2.0 was proposed, which is an intelligent tutoring system designed to help students in writing their papers, specifically in evaluating lexical richness in seven sections of the thesis.

Lexical richness refers to the diversity and depth of the vocabulary used. Additionally, a new feature is being added to the platform that allows for the identification of the structure of objectives using pre-trained BERT models. The evaluated sections of a thesis are the problem statement, objectives, justification, methodology, hypothesis, research questions, and conclusions, using measures of lexical richness such as variety, density, and sophistication.

Lexical capacity is the writer's ability to use vocabulary appropriately, it is a fundamental indicator of writing quality. Studies have shown that students with an adequate and varied vocabulary tend to excel in their studies. In the work [6], although it focuses on English as a second language, it discusses the importance of lexical richness, and it was found that lexical sophistication is the most influential factor contributing to higher quality in writing. The correlation analysis revealed that the use of lexical diversity, sophistication, and fluency affects the quality of writing and can manifest differently in a text depending on the various scoring ranges.

2   Related Work

Traditionally, the evaluation of university theses has been carried out by academic experts in the field of study. With technological advancements, various platforms like Turnitin and Grammarly have emerged that, although they focus on plagiarism detection or grammar correction, represent a step towards the automation of written assessments.

Until now, few systems have integrated metrics of lexical richness into their evaluations. Those that exist tend to be specialized tools for linguistic research rather than for educational assessment. Furthermore, consider that these tools are often designed for the English language, which further reduces the number of instruments available for the Spanish language. In the work done in [7], the importance and evaluation of lexical diversity in texts is addressed. Lexical diversity refers to the variety of words used in a text, serving as an indicator of lexical richness. This metric has found applications in fields such as stylistics, neuropathology, language acquisition, data mining, and forensic science. However, the challenge has been to find a robust index that is not sensitive to the length of the text.

Intelligent Tutoring Systems (ITS) have the potential to transform teaching and learning. Although a lot of effort has been invested in its design and development, mixed results have been reported regarding its effectiveness. The ITS are programs that offer personalized tutoring and can determine the learning path, select and recommend content, among other things. The systematic review in [8] revealed a complicated landscape regarding research on ITS using the social experimentation method between 2011 and 2022. Although information and communication technologies have the power to support teaching and learning, technology alone cannot guarantee its success. Contextual and social factors in real education can influence the observed effectiveness of intelligent tutoring systems.

The TURET 2.0 platform developed in [5] is a tool designed to help students enhance the lexical richness of their writings, thereby improving the quality of the final document. This allows academic advisors to focus more on the content of the thesis rather than on the structure or vocabulary. The goal is for it to be a motivating tool, not just another one to meet writing requirements. The formulas presented in [5] will carry out lexical richness evaluations and present the results to the user, providing a friendly interface with simultaneous analyses. In other words, as the student writes on our platform, the analysis of Lexical Richness will be carried out.

In recent years, BERT models (Bidirectional Encoder Representations from Transformers) have proven to be highly effective for various natural language processing (NLP) tasks, including text classification. BERT is based on a transformer architecture, which allows for bidirectional capture of the context of words in a sentence, significantly improving performance in NLP tasks [12]. A pre-trained BERT model named Multilingual-BERT (mBERT) [15,16] model with Spanish corpus was used for the classification task, with the aim of identifying the structure of the objectives. This approach allows for the precise identification of the relationships between different parts of the text, which is crucial for assessing coherence and cohesion in academic writing [13]. BERT has a wide range of applications in different languages [15]. The use of pre-trained BERT in Spanish has proven to improve accuracy in the task of identifying the parts of objectives due to its great classification capability, especially when compared to other more traditional classification methods. This is due to the model's ability to capture complex contexts and nuances of the Spanish language, which is essential for deep text analysis tasks.

3   Methodology

The purpose of the methodology presented in this section is to thoroughly explain the process of designing, developing, and implementing the web platform in order to enrich the vocabulary of the students.

3.1   Definition of Metrics and Variables

The evaluation of the seven sections of a thesis on the platform is based on three main metrics: lexical variety, lexical density, and lexical sophistication. The mentioned metrics not only reflect different aspects of language, but they also provide a comprehensive view of the quality and depth of the written content.

Before proceeding with the evaluation of the input document, it is essential to clearly establish the variables that will be used in the subsequent calculations.

Table 1. Definition of Variables

Variable name	Definition
N	Total number of tokens (words)
Nlex	Total number of words that add meaning to the text
Tlex	Total number of unique words that add meaning to the text
Nslex	Words within the 1000 most common words in Spanish (according to the Royal Spanish Academy)

In Table 1, the definitions of the fundamental concepts that allow for the determination of diversity, density, and complexity are presented. The values are calculated immediately in the web browser as the student inputs content into the platform. This approach reduces some of the server's load by delegating that task to the user's browser, and it also eliminates the communication latency between the server and the user (Table 2).

Table 2. Formulas for calculating lexical metrics

Metric	Formula
Lexical Variety	$Tlex/Nlex$
Lexical Density	$Tlex/N$
Lexical Sophistication	$Nslex/Nlex$

The first metric to consider is lexical variety, which is obtained by dividing the number of unique lexical terms by the total number of lexical terms. Lexical variety, as previously mentioned, refers to the breadth of vocabulary used in a text. The presence of a wide variety of terms indicates that the author has an extensive lexical repertoire and is capable of articulating concepts from various perspectives.

The following is the calculation of lexical density. This indicator represents the relationship between lexical words (such as nouns, verbs, adjectives, and adverbs) and the total number of words in the text (including stop words). A text with greater lexical richness tends to be more informative and precise, avoiding repetitions and unnecessary words. To determine this value, the division is made between the number of unique lexical terms and the total number of tokens or words present in the text, considering those that do not contribute to the meaning of the text (stop words).

Finally, the level of sophistication is assessed. For this purpose, we turn to the tool provided by the Royal Spanish Academy (RAE), which consists of a list of the 1000 most frequently used words in the Spanish language. To calculate this value, the number of common words according to the Royal Spanish Academy is divided by the total number of lexical terms, which are words that contribute meaning to the text and are analyzed by Freeling [17].

3.2 Analysis of Objectives

The platform also offers the possibility to analyze objectives through a deep learning architecture with M-BERT. In order to achieve this feature, a pre-trained BERT model was used, as described in [11]. The ability to transfer learning is one of the most notable features of BERT. Once BERT has been pre-trained on a vast text corpus, it is possible to adapt it to smaller datasets to carry out specific tasks, such as text classification or named entity recognition, with a significantly lower training effort.

With this feature, the platform is be able to establish a correlation with the thesis objective and answer three methodological questions. Q1, Q2, and Q3. To take advantage of the ability to detect linguistic patterns that indicate the presence of specific questions, a variant of BERT with multilingual features mBERT. Despite the proficiency of BERT models tailored for the Spanish language such as BETO, we have opted for the multilingual BERT, as it does not exhibit a significant disparity for our application. The research undertaken in [16] indicates that this model has strong performance was used that was specifically trained for sentence classification tasks. Despite the fact that BERT has been pre-trained on an extensive corpus of data, it needs to be fine-tuned to identify specific questions of interest. The corpus has been manually labeled with the purpose of being used in the training of the model for the classification task. At the end of the process, three models were developed in order to address each of the questions posed.

The BERT models were integrated into the project after being trained. An API was used with Python to carry out this analysis. The platform uses the three pre-trained BERT models to identify each question individually. That is to say, one model is responsible for detecting answers to questions Q1, Q2, and Q3.

3.3 Technologies Used for the Development of the Platform

The platform is based on a web architecture that integrates the responsiveness of a Single Page Application (SPA) developed with ReactJS, along with the performance of the Freeling server hosted on a Linux operating system. In order to ensure efficient data management, Appwrite is used as a backend as a service

(BaaS) platform, with its main logic hosted in an API created with Python and Flask.

User Interface. The user interface has been developed as a Single Page Application (SPA) using ReactJS. This ensures a smooth and responsive user experience, where the page does not need to be completely reloaded when interacting with it. An additional advantage of this type of application is the ability to efficiently transfer certain calculations to the browser, avoiding server overload and eliminating latency between the client and the server. In the text editor, highlighted words provide instant feedback, allowing writers to immediately identify and correct areas for improvement. Additionally, the presentation of the results is done with an intuitive and aesthetically pleasing design, which simplifies the understanding of the analyses.

Freeling API. This application programming interface (API) is essential for the linguistic analysis of the text entered by the user, is hosted on a Linux server. The API processes the text, including the root words, which is essential for the linguistic evaluations that will be conducted later. The API in question is intended solely for use within this platform in order to ensure optimal response time.

Database and User Management. Appwrite has been selected as the preferred solution for database and user management, thanks to its robustness, advanced features, free availability, and strong community support. This backend service is responsible for: management of user permissions, storage of evaluation history and manage user registration and authentication.

Bridge API. The bridge API, created using the Python programming language and the Flask framework, acts as an intermediary between the user interface, the Freeling API, and the Appwrite service used by the platform. The responsibility involves processing requests from the user interface, establishing interactions with the database and the Freeling API, and then sending the processed results back to the user using an SSL certificate to encrypt the information. With the combination of these technologies, the user not only benefits from a real-time evaluation of their content as they write, but they can also enjoy an interactive and dynamic experience.

4   Results

This section graphically displays the final product of this work. The images display the user interface and the presentation of information once the entered text has been processed and analyzed.

In Fig. 1, the lexical richness analysis interface of the platform is shown. In this space, users have the opportunity to input the text they wish to analyze.

The highlighted words in yellow belong to terms that are among the 1000 most frequent words according to the Royal Spanish Academy. In contrast, the words highlighted with a red stroke indicate terms that have been repeated in the text. The results are displayed in the right section indicating the evaluation of each metric. To view the interface, click on the following link[1].

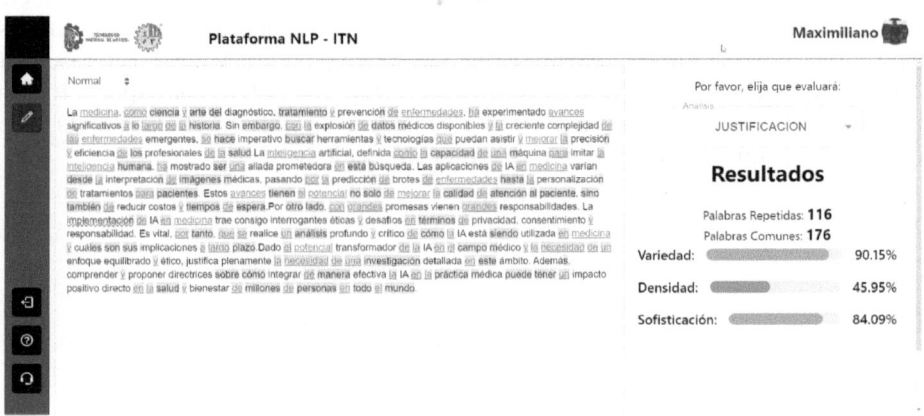

Fig. 1. User interface for the evaluation of lexical richness

In the interface presented in Fig. 2, the user can easily conduct objective analysis. The user writes or pastes the project's objective, and the platform will display the results, indicating the answer to each of the three questions.

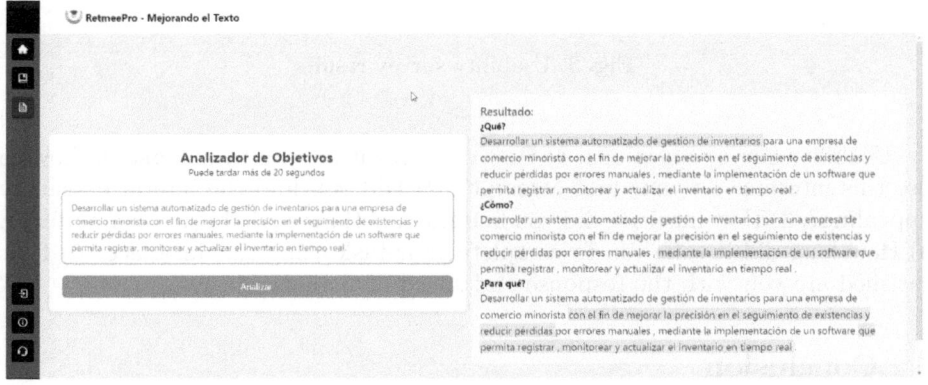

Fig. 2. Analysis of objectives on the platform

---

[1] Visit http://retmeepro.turet.com.mx ReetmePro, the final product of the current work.

4.1   Usability Survey

According to 17 user surveys, most users found the RetmeePro platform responsive and effective. The scale used for the questions was 1 to 10, with 10 being
a positive result. Some users reported problems when using the platform on
mobile devices, citing difficulties with the devices, mentioning difficulties with
the side menu, and visualization problems. This aspect is critical as it affects
the user experience and usability of the platform in mobile contexts. The results
allowed us to make improvements to the platform. The following is the result of
the survey in Fig. 3. The graph above shows that the colors correspond to the
five questions asked to the users, the x-axis corresponds to the students who
answered the survey, and the y-axis shows the scale used.

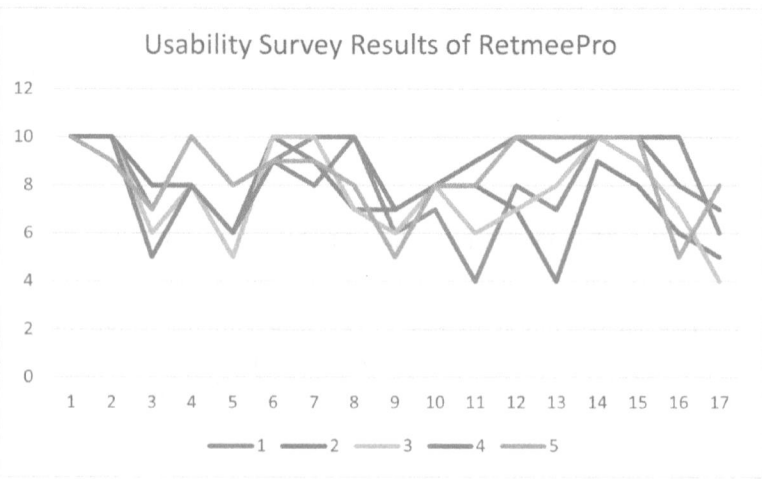

Fig. 3. Usability survey results

Usability survey questions: 1) How easy was it for you to understand how to
use this interface on your first attempt?, 2) Did you find the interface visually
appealing?, 3) How intuitive did you find the functions available in the interface?,
4) How long did it take you to complete your task using this interface? 5) How
satisfied are you with the response and speed of the interface?.

5   Conclusion

Writing a thesis is a fundamental activity in the academic training of university
students. However, the quality of writing and the expansion of vocabulary often
pose challenges. In this regard, the proposal is to present an online platform
created to assist in the linguistic analysis of theses through metrics that evaluate
vocabulary breadth. The platform provides real-time assessment to users by

combining technologies such as ReactJS and Python, along with the integration of the Freeling API. This allows them to receive instant feedback on their content. Understanding and learning are facilitated by visual cues and results presented in an accessible manner.

In further work, incorporating supplementary metrics or implementing more sophisticated machine-learning tools could aid the students. Integrating a more sophisticated feedback method by implementing a dialogue system would be feasible, as discussed in [10]. Some systems have developed the combination of dialogue generation with pedagogical strategies to assist students in their learning process. The platform will become a fundamental tool for students and researchers, improving the quality of thesis writing and significantly contributing to the academic field. Writing a thesis is an essential activity in the academic training of university students.

References

1. Schcolnik, M.: Digital Tools in Academic Writing? J. Acad. Writ. **8**(1), 121–130 (2018)
2. Manal, A.: Academic writing: challenges and potential solutions. Arab World Engl. J. (AWEJ) (6) (2020)
3. Dale, R., Viethen, J.: The automated writing assistance landscape in 2021. Nat. Lang. Eng. **27**, 511–518 (2021)
4. Ippolito, D., Yuan, A., Coenen, A., Burnam J.S.: Creative writing with an AI-powered writing assistant: perspectives from professional writers (2022)
5. González-López, S., López-López, A.: Lexical analysis of student research drafts in computing. Comput. Appl. Eng. Educ. **23**, 638–644 (2015). https://doi.org/10.1002/cae.21638
6. Seung, H.: Lexical richness in EFL undergraduate students' academic writing. Engl. Teach. **74**(3), 3–28 (2019)
7. McCarthy, P.M., Jarvis, S.: MTLD, vocd-D, and HD-D: a validation study of sophisticated approaches to lexical diversity assessment. Behav. Res. Methods **42**, 381–392 (2010)
8. Wang, H., Tlili, A., Huang, R., et al.: Examining the applications of intelligent tutoring systems in real educational contexts: a systematic literature review from the social experiment perspective. Educ. Inf. Technol. **28**, 9113–9148 (2023)
9. Van Hout, R., Vermeer, A.: Comparing measures of lexical richness. Model. Assess. Vocabulary Knowl. **93**, 115 (2007)
10. Paladines, J., Ramirez, J.: A systematic literature review of intelligent tutoring systems with dialogue in natural language. IEEE Access **8**, 164246–164267 (2020)
11. González-López, S., Montes-Rosales, Z.G., López-Monroy, A.P., López-López, A., García-Gorrostieta, J.M.: Short answer detection for open questions: a sequence labeling approach with deep learning models. Mathematics **10**, 2259 (2022)
12. Devlin, J., Chang, M., Lee, K., Toutanova, K.: BERT: pre-training of deep bidirectional transformers for language understanding. arXiv:1810.04805 (2019)
13. Cañete, J., Donoso, S., Bravo-Marquez, F., Carvallo, A., Araujo, V.: ALBETO and DistilBETO: lightweight spanish language models. arXiv:2204.09145 (2023)
14. Cañete, J., Chaperon, G., Fuentes, R., Ho, J., Kang, H., Pérez, J.: Spanish pretrained BERT model and evaluation data. arXiv:2308.02976 (2023)

15. Pires, T., Schlinger, E., Garrette, D.: How multilingual is Multilingual BERT?. arXiv:1906.01502 (2019)
16. Gonzalez-Lopez, S., Bethard, S.: Transformer-based cynical expression detection in a corpus of Spanish YouTube reviews. In: Barnes, J., De Clercq, O., Klinger, R. (eds.) Proceedings of the 13th Workshop on Computational Approaches to Subjectivity, Sentiment, & Social Media Analysis, Toronto, Canada, pp. 194–201. Association for Computational Linguistics (2023). https://doi.org/10.18653/v1/2023.wassa-1.18, https://aclanthology.org/2023.wassa-1.18
17. Padró, L., Stanilovsky, E.: FreeLing 3.0: towards wider multilinguality. In: Proceedings of the Language Resources and Evaluation Conference (LREC 2012), Istanbul, Turkey. ELRA (2012)

Competence-Based Student Modelling with Dynamic Bayesian Networks

Rafael Morales[1][(✉)] and L. Enrique Sucar[2]

[1] University of Guadalajara, Guadalajara, Jalisco, Mexico
`rmorales@suv.udg.mx`
[2] National Institute for Astrophysics, Optics and Electronics,
San Andrés Cholula, Mexico
`esucar@inaoep.mx`

Abstract. We present a general method for using a competences map, with generalization/specialization and inclusion/part-of relationships between competences, in order to build an overlay student model in the form of a dynamic Bayesian network in which conditional probability distributions are defined per relationship type. The method is demonstrated using a sample competences map for tracing the development of the corresponding competences by three hypothetical students exhibiting different performances along an online course (low to medium performance, medium to high performance but with low final score, and two terms medium to high performance). The results obtained suggest that the proposed way for constructing dynamic Bayesian student models on the basis of competences maps could be useful to monitor competence development by real students in online courses.

Keywords: Student modelling · dynamic Bayesian network · competence · competences map

1 Introduction

Competences have grown in popularity in the educational world [8,13,18,19], and so the interest on developing computational models that can be used to support a variety of educational processes, including the monitoring of competence development by students. Research in this area is important because little information is usually available regarding what competences students have developed along their studies; beyond the learning objectives of their educational programmes, the titles of the courses they have taken, and their final scores, little is known concerning student performance in a variety of tasks, and internal characteristics such as knowledge, skills, attitudes, and values.

As e-learning facilitates automatic harvesting of information regarding student behaviour and performance, we have proposed to provide the digital e-learning environment with detailed information regarding competences, their interrelationships, and their relations to course activities, so that student performance in the latter can be used as evidence about the current state of their

L. Martínez-Villaseñor et al. (Eds.): MICAI 2024 Workshops, LNAI 15465, pp. 25–36, 2025.
https://doi.org/10.1007/978-3-031-83882-8_3

competences, and then use it to support educational processes [15]. We have proposed generic mechanisms for creating probabilistic graphical models to trace the development of competences on the basis of competences maps [16].

We present here initial results from using such mechanisms for building an overlay student model in the form of a dynamic Bayesian network, and using it to trace the development of competences by three hypothetical students exhibiting different performances along courses that broadly resemble those that can be found in e-learning, and comparing its estimates against those provided by a sample of teachers. The results suggest correlations among both sets of estimates, so the model is operational, yet the teachers seemed to be more optimistic and certain about the development of competences by students.

2   Related Work

Competences are demonstrated through performance on tasks in certain kinds of situations, yet it is assumed they require the mobilization of both personal attributes and external resources. In any case, a competence acts as a whole and cannot be assessed using a check list for personal attributes, but through evaluation of the performance as a whole [10].

There has been a considerable amount of work on computational representations for competences in this century: on a standard [2] and a recommendation [5] on how to encode competences for exchange between applications, including description of competences in terms of their components and their relationships; on extending the recommendation to have better descriptions of relationships between competences, and ways for encoding levels of competence development [22]; on providing detailed descriptions of competence elements, including composition and specialization relationships, and procedures structures for complex descriptions of competences [20] and computational tools to deal with them [23]; on developing rather formal definitions of competences for knowledge management inside an organization, amenable of both graphical and symbolical representations [21]; on extending the recommendation to include relationships among competences as part of the model, distinguishing between general and specific competences, in the sense of performances in domains and subdomains [7].

Regarding the use of Bayesian probability estimation and Bayesian networks (both static and dynamic) for student modelling, it has been explored at least since the mid-1990s [4,6,9,12,14,28]. A study [3] suggests that it has been motivated by the large amount of uncertainty in estimating the cognitive or affective state of students, their sound foundations in probability theory to make inferences, and their transparency compared to other numerical representations such as neural networks. However, the task of defining conditional probability distributions can become daunting as the network grows, and usually demands the availability of a large amount of data and the use of machine learning techniques [12,25]. Recent work [12] demonstrates a generic way of building student models as dynamic Bayesian networks on the basis of "skill topologies", defined by prerequisite relationships between skills, and a large collection of data.

The main contributions of this work are the integration of two fields of research that have been explored separately, computational representations of competences maps and (dynamic) Bayesian student modelling, and the proposal of a method for defining conditional probability distributions on the basis of types of relationships in competences maps beyond the prerequisite type, and without the need to previously accumulate large amounts of data.

3 Competence Maps

Our work is based on the notion of *competence* as the ability to carry out a given action in a given context through the mobilization of various cognitive, affective and conative resources, such as knowledge, skills, attitudes and values [1]. This definition allow us to define two kinds of relationships between competences: *generalization/specialization*, generated through the removal or addition of resources (we call them *competence attributes*), respectively, and *inclusion/part-of*, generated by considering the attributes of some competences being part of a larger one, or by distributing the resources of a given competence along some sub-competences. We call a collection of competences, each one defined by the action to be carried out and the internal attributes it demands (knowledge, skills, attitudes, and values), interrelated by relationships of generalization/specialization and inclusion/part-of, a *competences map*.

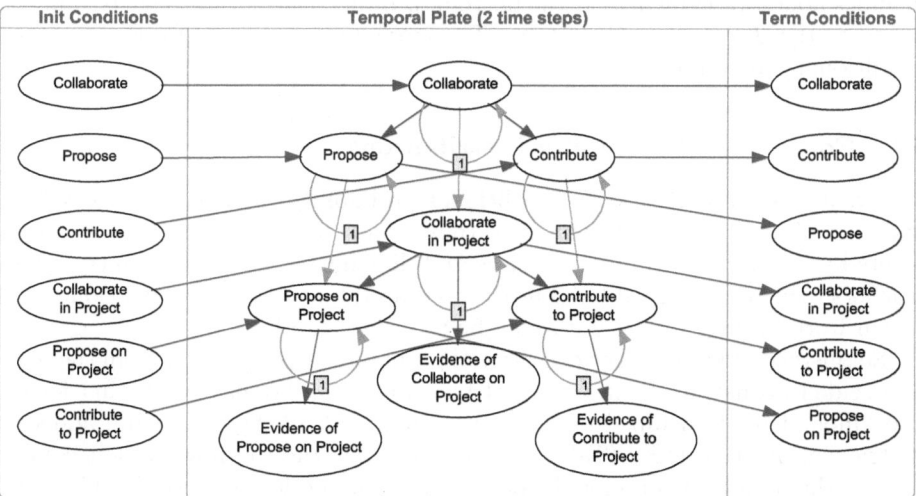

Fig. 1. Example of a dynamic Bayesian network corresponding to a competences map. Straight solid arrows in the *Temporal Plate* correspond to inclusion/part-of relationships, whereas straight dashed arrows correspond to generalization/specialization relationships in the competences map.

4   Dynamic Bayesian Networks

A competence can be observed only through performances in concrete situations, and such performances are considered evidence of the level of competence. The development of a more specific competence is evidence of the development of a more general one—as they share some key attributes—and the development of a super-competence cannot occur independently of its sub-competences. So, in the case of competences, we may attribute some kind of causality to the generalization/specialization and inclusion/part-of kinds of relationships. Hence, we propose to create overlay student models [11, 26] to trace the development of competences by associating a belief on the degree of development to each competence in the map, transforming the competences map into a Bayesian network [24]. As student competences evolve along time, on the basis of their previous state and new learning experiences, beliefs about their current levels of development are dependent both on beliefs about their previous state and new evidence, so the Bayesian network should be dynamic.

A dynamic Bayesian network (DBN) corresponding to a competences map is shown in Fig. 1. *Init(ial) Conditions* nodes are set to beliefs about the previous state of the competences while nodes in *Term(inal) Conditions* are used to recover the beliefs on the current state of the competences. The nodes in the *Temporal Plate* actually represent two instances of the (non-dynamic) Bayesian network built from the competences map. The first instance is linked to the *Init(ial) Conditions* and contains previous beliefs, whereas the second instance includes nodes for evidences at the bottom, produces the current beliefs, and it is linked to the *Term(inal) Conditions*. The curved arrows stand for the temporal conditional probabilities.

4.1   Conditional Probability Distributions

Today, it is common to learn conditional probability distributions from large amounts of data [12,17]. In this case they would be about competence assessments by teachers, but are both scarce and difficult to get in our country. Many teachers have been trained for competence-based teaching and evaluation, yet it is still work in progress, so data may be too coarse and dirty.

So, instead of learning the conditional probability distributions from the data, we decide to construct them from first principles [16]. We have done that for the two kinds of relationships in competences maps (Table 1), plus those added in its translations to a DBN (Table 2)—in each table, the top row and left column include the possible values for the parent and child node, respectively. The fuzzy terms in the tables are used to describe degrees of probability, which need to be instantiated to specific numerical values (Table 3).

From a social constructivist perspective [27], we decided to move away from typical binary variables (e.g., mastered or not mastered, as in [6]) and to have three possible values for competence development, representing that the student cannot perform the associated activity, even with scaffolding (*Low*), the student can perform the associated activity, but only with scaffolding (*Medium*), and the student can perform the associated activity on their own (*High*).

Concerning the design of the DBN, the final decision was on which numerical values to assign to the fuzzy terms used to describe the conditional probability distributions. Given the rate of decay of some probabilities in the conditional probability distribution for the inclusion/part-of relationship (Table 1), we decided to give a similar behaviour to the fuzzy terms, so that the numerical values are those shown in Table 3.

Table 1. Conditional probability distribution for the inclusion/part-of and specialization/generalization relationships. The variable n stands for the number of sub-competences. Source [16].

	Inclusion/part-of			Specialization/generalization		
	Low	Medium	High	Low	Medium	High
Low	$\dfrac{3^{n-1}}{3^n - 2^n}$	$\dfrac{1}{2^n}$	Very small	Large	Medium	Small
Medium	$\dfrac{3^{n-1} - 2^{n-1}}{3^n - 2^n}$	$\dfrac{1}{2}$	Small	Small	Large	Large
High	$\dfrac{3^{n-1} - 2^{n-1}}{3^n - 2^n}$	$\dfrac{2^{n-1}-1}{2^n}$	Large	Very small	Small	Medium

Table 2. Conditional probability distribution for competence/evidence and past/present relationships.

	Competence/evidence			Past/present		
	Low	Medium	High	Low	Medium	High
Low	Large	Medium	Small	Large	Very small	Tiny
Medium	Small	Large	Medium	Very small	Large	Very small
High	Very small	Medium	Large	Tiny	Tiny	Large

Table 3. Numerical values associated to the fuzzy terms used to describe the conditional probability distributions.

Term	σ	Value
Large	0	0.5
Medium	-1	0.15865525393145707
Small	-2	0.02275013194817919
Very small	-3	0.00134989803163009
Tiny	-4	0.00003167124183311

5   Simulations of Students and Courses

We have simulated two courses: one devoted to the development of the specialized competences included in the map shown in Fig. 1, and another with some activities that make use of them; they run for fourteen weeks, plus two weeks for revisions and additional examinations, and five weeks of holidays. We have also simulated students exhibiting three different performances: low to medium performance, medium to high performance but with final failure, and two terms medium to high performance. These performances are represented by the sequences of evidences shown in Table 4, where the second column shows the competence to which the performance is related: to *propose* on project, to *contribute* to project, or to *collaborate* in project; and columns third to fifth include the level of performance observed, with the levels Low, Medium, and High translated to numbers (0, 1, and 2, respectively). The time unit is a week, so the dynamic Bayesian network is updated fourteen times along the first course, for all three students. For the third student, the network is updated another seven times, to account for the two revision weeks and the five weeks of holidays, and then updated fourteen times more, along the second course.

Table 4. Evidences for simulated students: low to medium performance (L2M), medium to high performance (M2H), but with final product missing, and two terms medium to high performance (LT M2H)).

Week	Competence	L2H	M2H	LT M2H
1	Propose	0	1	1
2	Contribute	0	1	1
4	Propose	1	2	2
5	Contribute	0	1	1
7	Collaborate	0	1	1
10	Propose	1	2	2
11	Contribute	1	2	2
14	Collaborate	1	0	2
23	Propose			2
24	Contribute			2
25	Collaborate			2
35	Collaborate			2

6   Comparison with Estimations by Teachers

In order to evaluate the overall design of our system, we decided to compare the estimations it produces against estimations provided by teachers. We designed a questionnaire in which we ask teachers to estimate the levels of the more specific competences of our simulated students (given the evidence in time slices 3, 6, 7, 12, 14, and 35) and to provide a degree of certainty for their estimations. In addition, we ask them to estimate the levels of the more general competences at the end of the course and their degree of certainty on that. We ran an online poll among colleagues and doctoral students, 20 of which answered it: 65% said they are full time teachers, 25% subject teachers, 5% mainly researchers, and another 5% do not teach. 60% of the participants said that they teach primarily at the undergraduate level, 35% said that they are mainly postgraduate teachers, and only 5% classify themselves primarily as high school teachers.

The teachers provided their estimations for the competence levels using a Likert scale with five values: (0) Low, (0.5) Rather Low, (1) Medium, (1.5) Rather High, and (2) High. They provided their degrees of certainty in a similar scale. We ran the Shapiro-Wilk test of normality on the estimations of competence levels provided by the teachers, and we found that most distributions are far from normal, so we proceeded to the analysis of results using non parametric methods, which suggest a high consistency among them.

Then we compared the responses provided by the teachers with the estimations provided by the system. We noticed that although the latter are, in general, lower than the median of the former (Fig. 2, Fig. 3, and Fig. 4), there seems to be a correlation between them, so we calculated their Pearson coefficient of correlation, and the results are shown in Table 5. They suggest that both teachers and system followed the same pattern in estimating competence levels when there was more or less frequent evidence and it indicated an unsurprising behaviour, but the differences grow when there are surprises and when there are fewer evidences. We calculated the first and third quartiles among the estimations provided by the teachers and we compared them against the estimations generated by the system. The results are that, in general, the estimations provided by the system are out of the inter-quartile range. These results confirm the significance of differences in estimations by the teachers and the system.

Uncertainty in the beliefs maintained by the system shows relatively high sensitivity to changes in the evidence, whereas uncertainty in estimations by the teachers seems almost stable. The comparison of estimations of the final state of development of the most generic competences, and the corresponding uncertainties, by the teachers and the system, shown in Table 6, suggest a similar pattern of the teachers being more optimistic and certain at providing higher estimations of competence development.

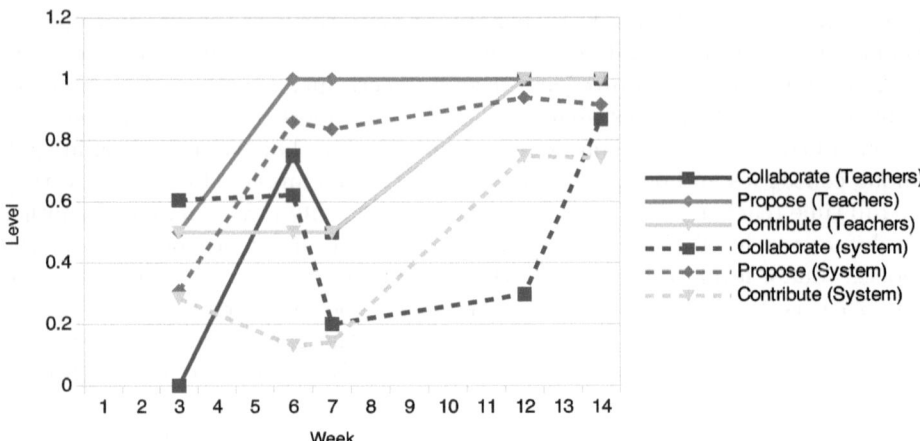

Fig. 2. Comparison of estimates by teachers and the system of competence levels of the student with low to medium performance.

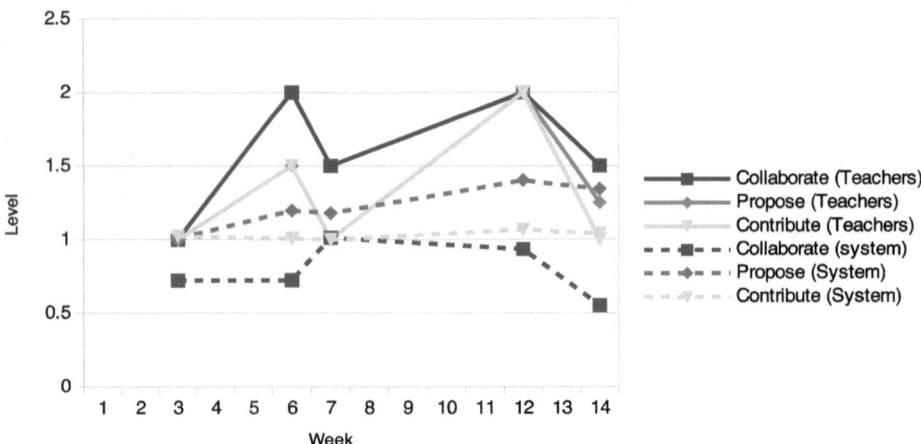

Fig. 3. Comparison of estimates by teachers and the system of competence levels of the student with medium to high performance, who failed on the last evaluation.

Table 5. Pearson correlation coefficients between competence level estimations by the teachers and the system for students of low to medium performance (L2M), medium to high but failing on final evaluation (M2H), and long term steady medium to high (LT M2H).

	Competence level		
	Collaborate in project	Propose on project	Contribute to project
L2M	0.1026	0.9871	0.9808
M2H	0.2385	0.7406	0.6307
LT M2H	0.3458	0.8159	0.8512

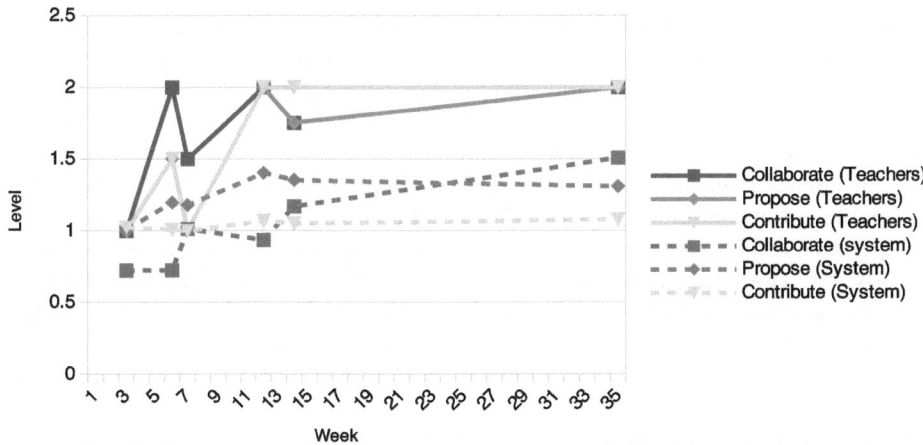

Fig. 4. Comparison of estimates by teachers and the system of competence levels of the student with medium to high performance along tow periods.

Table 6. Comparison of estimates and incertainties as provided by the teachers and the system regarding the development of the most generic competences—to collaborate (Col), to propose (Prop), and to contribute (Cont)—by students of low to medium performance (L2M), medium to high performance with failure in final evaluation (M2H), and steady medium to high performance along two terms (LT M2H).

(Median)	L2M			M2H			LT M2H		
	Col	Prop	Cont	Col	Prop	Cont	Col	Prop	Cont
Teacher estimation	1	1	1	1.5	1.5	1.25	2	2	1.75
Teacher uncertainty	0.25	0.25	0.25	0.25	0.25	0.25	0.25	0.25	0.25
System estimation	1.01	1.32	1.25	0.97	1.33	1.32	1.05	1.40	1.39
System uncertainty	1	0.93	0.96	1	0.92	0.93	0.99	0.87	0.88

7 Conclusions

We have shown a way of creating overlay student models as dynamic Bayesian networks built on top of competences maps including generalization/specialization and inclusion/part-of relationships. It works by defining a conditional probability distribution per type of relationship (and cardinality, in the case of inclusion/part-of relationships), so it can be applied to any map restricted to those common types of relationships. This approach provides a general way for assigning weights to the relationships between competences, which does not rely in large amounts of data, and considers competences very much as holistic entities, but by their decomposition into sub-competences. In that sense, it is quite different from fine grained approaches criticized by [10], and it would be much easier to make it work on real life conditions, provided it delivers reasonable results.

Our approach takes advantage of the functionality of Bayesian networks to propagate the effect of concrete evidence on a competence level, to beliefs on competences related to it in map, allowing for indirect evidence to fine-tune beliefs in competence development—an important difference from earlier models such as Bayesian Knowledge Tracing (BKT) [6]. The results obtained from implementing this method on a sample competences map, performing the modelling of competence development by some simulated students, and then comparing the competence levels estimated by the prototype against those estimated by teachers, suggest that the inferencing carried out by the dynamic Bayesian network goes along what teachers estimate from evidences of performance by students. However, teachers seem to be significantly more optimistic, and confident, on competence development by students, particularly when they get evidence of good performance.

Regarding the limitations of our work, we have assumed all evidences to be hard ones—e.g. no cheating—from what we consider common, yet nevertheless invented, student performances. Regarding the participants in our study, their task was not what they are used to do: a kind of meta-analysis on the evidence produced by others, instead of directly observing the performance of students and then producing the evidence themselves—what it would be their actual task if the system gets implemented for some educational programme.

Future work includes fine tuning of our conditional probability distributions, expanding the development and testing with evidence produced from real (historical) data. In the long run, we expect to do real-time, real-life student modelling, considering not only information provided by teachers, but information gathered from other sources, inside and outside schools.

Acknowledgement. The core of our implementations is based on the SMILE reasoning engine for graphical probabilistic models, while images of them included in this paper were created using the GeNIe Modeler, both available free of charge for academic research and teaching use from BayesFusion, LLC. This work was supported by the Common Space for Distance Higher Education (ECOESAD) and the Program for Teachers Professional Development (PRODEP).

Disclosure of Interests. The authors have no competing interests to declare that are relevant to the content of this article.

References

1. Chan, M.E., González, S.C., Morales, R.: A competency analyser as a knowledge-based approach for making e-learning more flexible and personalised. In: EDULEARN10 Proceedings CD. 2nd International Conference on Education and New Learning Technologies (Edulearn), Barcelona, pp. 1607–1612 (2010)
2. Competency Data Working Group: 1484.20.1-2007 IEEE Standard for Learning Technology–Data Model for Reusable Competency Definitions. IEEE Standard for Learning Technology, The Institute of Electrical and Electronics Engineers, Inc., New York (2008)

3. Conati, C.: Bayesian student modeling. In: Advances in Intelligent Tutoring Systems. Studies in Computational Intelligence, vol. 308, pp. 281–299. Springer (2010)
4. Conati, C., Gertner, A., VanLehn, K.: Using Bayesian networks to manage uncertainty in student modeling. **12**(4), 371–417 (2002)
5. Consortium, I.G.L.: IMS reusable definition of competency or educational objective specification. Technical report Version 1.0 Final Specification, IMS Global Learning Consortium, Inc. (2002)
6. Corbett, A.T., Anderson, J.R.: Knowledge tracing: modeling the acquisition of procedural knowledge. **4**(4), 253–278 (1995)
7. El Asame, M., Wakrim, M.: Towards a competency model: a review of the literature and the competency standards. Educ. Inf. Technol. **23**(1), 225–236 (2018). https://doi.org/10.1007/s10639-017-9596-z
8. Gordon, J., et al.: Key competences in Europe: opening doors for lifelong learners across the school curriculum and teachers education. Technical report 87/2009, CASE – Center for Social and Economic Research, Warsaw, Poland (2009)
9. Greer, J.E., Zapata-Rivera, J.D., Ong-Scutchings, C., Cooke, J.E.: Visualization of Bayesian learner models, Le Mans, France, pp. 9–13 (1999)
10. Hager, P., Gonczi, A.: What is competence? Med. Teach. **18**(1), 15–18 (1996). https://doi.org/10.3109/01421599609040255
11. Holt, P., Dubs, S., Jones, M., Greer, J.: The state of student modelling. In: Greer, J., McCalla, G.I. (eds.) Student Modelling: The Key to Individualized Knowledge-Based Instruction. NATO ASI Series, vol. 125, pp. 3–35. Springer (1994)
12. Käser, T., Klingler, S., Schwing, A.G., Gross, M.: Dynamic Bayesian networks for student modeling. IEEE Trans. Learn. Technol. **10**(4), 450–462 (2017). https://doi.org/10.1109/TLT.2017.2689017
13. Lurie, H., Garrett, R.: Deconstructing competency-based education: an assessment of institutional activity, goals, and challenges in higher education. J. Competence-Based Educ. **2**(3), 1–19 (2017). https://doi.org/10.1002/cbe2.1047
14. Millan, E., Loboda, T., Luis Perez-de-la-Cruz, J.: Bayesian networks for student model engineering. Comput. Educ. **55**(4), 1663–1683 (2010). https://doi.org/10.1016/j.compedu.2010.07.010
15. Morales, R.: Towards an intelligent environment for distance learning. World J. Educ. Technol. **1**(2), 110–117 (2009)
16. Morales-Gamboa, R., Sucar-Succar, E., Ruiz-Hernández, E., Chan-Núñez, M.E., González Flores, S.C.: Probabilistic relational learner models based on competence maps. Res. Comput. Sci. **146**, 77–86 (2017)
17. Murphy, K.P.: Dynamic Bayesian networks: representation, inference and learning. Ph.D. thesis, University of California at Berkeley (2002)
18. Núñez Cortés, J.A.: El modelo competencial y la competencia comunicativa en la educación superior en América Latina. Foro Educ. **14**(20), 467–488 (2016). https://doi.org/10.14516/fde.2016.014.020.023
19. OECD: The Future of Education and Skills: Education 2030. Technical report, OECD (2018)
20. Paquette, G., Léonard, M., Lundgren-Cayrol, K., Mihaila, S., Gareau, D.: Learning design based on graphical knowledge-modelling. Educ. Technol. Soc. **9**(1), 97–112 (2006)
21. Pépiot, G., Cheikhrouhou, N., Furbringer, J.M., Glardon, R.: UECML: unified enterprise competence modelling language. Comput. Ind. **58**, 130–142 (2007). https://doi.org/10.1016/j.compind.2006.09.010

22. Sampson, D., Karampiperis, P., Fytros, D.: Developing a common metadata model for competencies description. Interact. Learn. Environ. **15**(2), 137–150 (2007). https://doi.org/10.1080/10494820701343645
23. Stoof, A., Martens, R.L., van Merriënboer, J.J.G.: Web-based support for constructing competence maps: design and formative evaluation. Education Tech. Research Dev. **55**(4), 347–368 (2007). https://doi.org/10.1007/s11423-006-9014-5
24. Sucar, L.E.: Bayesian networks: representation and inference. In: Probabilistic Graphical Models: Principles and Applications. Advances in Computer Vision and Pattern Recognition, 2 edn., pp. 111–151, Springer, Cham (2021). https://doi.org/10.1007/978-3-030-61943
25. Sucar, L.E.: Probabilistic Graphical Models: Principles and Applications. Advances in Computer Vision and Pattern Recognition, 2 edn. Springer, Cham (2021). https://doi.org/10.1007/978-3-030-61943
26. VanLehn, K.: Student modelling. In: Polson, M.C., Richardson, J.J. (eds.) Foundations of Intelligent Tutoring Systems, pp. 55–78. Lawrence Erlbaum Associates, New Jersey (1988)
27. Vygotsky, L.S., Cole, M.: Mind in Society: The Development of Higher Psychological Processes. Harvard University Press (1978)
28. Zapata-Rivera, J.D., Greer, J.E.: Inspecting and visualizing distributed bayesian student models, pp. 544–553. Springer (2000)

XploRe: XR Tool for Learning About the Solar System and Its Physical Phenomena

José Miguel Gil-Núñez[1] (ID), María Blanca Ibañez-Espiga[2] (ID),
Ramón Zataraín-Cabada[1] (ID), and María Lucía Barrón-Estrada[1](✉) (ID)

[1] Tecnológico Nacional de México Campus Culiacán, Culiacán, Sinaloa, México
{jose.gn,ramon.zc,lucia.be}@culiacan.tecnm.mx
[2] Universidad Carlos III de Madrid, Leganés, Madrid, Spain
mbibanez@it.uc3m.es

Abstract. This article presents the development and evaluation of XploRe, an Extended Reality tool designed to enhance the learning experience of secondary school students, specifically focusing on the Solar System and its associated physical phenomena. XploRe integrates WebXR technology, ensuring broad compatibility across various devices such as PCs and mobile platforms, and supports multiple viewing modes, including Web, Augmented Reality (AR), and Virtual Reality (VR).

To assess the effectiveness of XploRe, a series of controlled experiments were conducted with secondary school students. The evaluation process included a pre-test to evaluate students' knowledge about the topics, interaction with the experiments within XploRe, and subsequent post-tests to measure learning outcomes.

Keywords: Solar System · Gravity · Extended Reality · Intelligent Learning Environment

1 Introduction

The learning of STEM (Science, Technology, Engineering, and Mathematics) subjects is critically important from an early stage of cognitive development, as it helps to cultivate essential skills such as critical thinking and problem-solving. When these subjects are complemented with gamification and game-based learning, children and adolescents can assimilate knowledge more effectively [1] (Sydon & Phuntsho, 2021).

Learning environments are powerful tools that support the teaching of various subjects to a large number of students, including those interested in STEM-related courses. These environments facilitate continuous interaction and communication, enhancing knowledge construction while also allowing for the early identification of students who may be at risk of academic failure [2] (Alves, Miranda, & Morais, 2017). Additionally, Extended Reality (XR) tools, such as immersive Virtual Reality (VR) and Augmented Reality (AR), offer a wide range of educational benefits, from improving interactivity

L. Martínez-Villaseñor et al. (Eds.): MICAI 2024 Workshops, LNAI 15465, pp. 37–44, 2025.
https://doi.org/10.1007/978-3-031-83882-8_4

and motivation to providing safe and accessible environments for learning and professional training, as well as supporting the education of individuals with disabilities [3] (Freina & Michella, 2015). Furthermore, learning environments can integrate Bloom's Taxonomy in the creation of exercises presented to students within the system. Using Bloom's Taxonomy allows for the division of learning topics into different levels, which involves a gradual increase in the depth and complexity of exercises, thereby enhancing students' comprehension as they progress through the learning environment [4] (Huitt, 2011). The main contribution of this article is the integration of extended reality technologies such as BABYLONjs and the WebXR API within a learning environment using the Bloom Taxonomy that can be accessed via any device compatible with AR and VR in a web app.

This article includes a summary of previous works in the area, the methodology to develop XploRe, the main interfaces of the web application, the process of evaluation of the tool, a statistical analysis of the results, and a discussion with conclusions.

2 Related Work

Many previous studies have explored the use of XR in educational settings, particularly in subjects related to learning of the Solar System. For example, the "Planetland" application utilized XR to teach natural sciences to elementary students, demonstrating significant improvements in student engagement and understanding [5]. Similarly, "SpaceVR" used VR to immerse students in space-related content, such as solar flares and satellite repairs, showing positive results in both learning outcomes and user satisfaction [6]. Rejón et al. (2023) created an AR-based application for fifth and sixth graders, utilizing 3D models of planets and the sun, integrated with Unity and Vuforia, to improve the understanding of astronomical concepts through interactive visuals [7]. Meanwhile, Tresnawati et al. (2019) designed an AR-based Solar System learning tool for primary students, which uses marker cards to project 3D animations of planets in real-time, enhancing engagement and comprehension [8]. Huang (2024) introduced a gesture-controlled VR system for exploring celestial bodies, offering students an immersive way to interact with 3D models and comprehend complex astronomical phenomena like planetary orbits and gravitational interactions, using XR Hands and OpenXR technology for gesture recognition [9].

However, many of these applications, while innovative, have limitations in terms of user interaction and customization. The lack of individualized learning experiences can result in a one-size-fits-all approach that may not address the specific needs of each student. XploRe aims to address these gaps by incorporating a more interactive and student-centered design, allowing for a more personalized learning experience.

3 Methodology

The design of the software for XploRe uses a hybrid architecture, consisting of a Client-Server architecture [10] and a layered architecture using the Model-View-Controller (MVC) pattern [11] (see Fig. 1).

XploRe uses a hybrid architecture where the MVC layers are split between the client and the server. The client manages the presentation layer and the business logic layer. It handles displaying the web application through a graphical user interface (GUI) and allows users to interact with and execute experiments and tests. The server, on the other hand, contains the data layer, which manages the Firebase database used to store user data and interactions with the experiments.

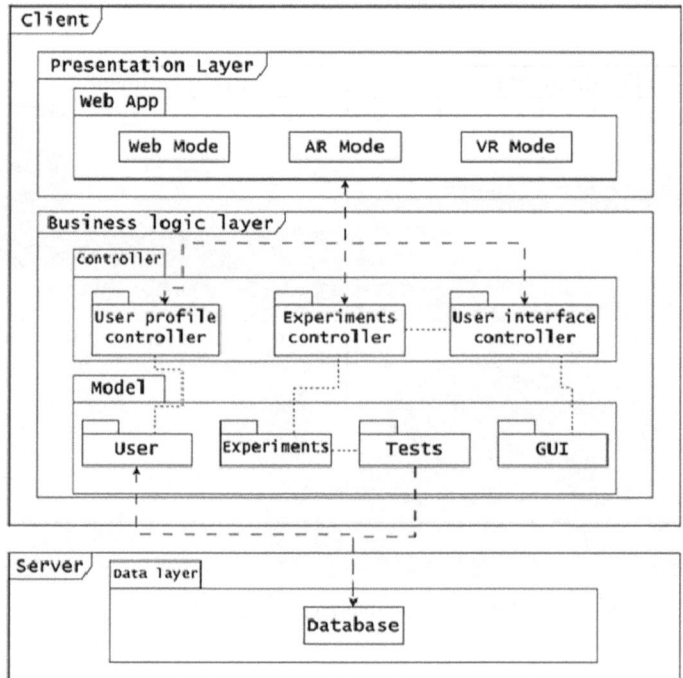

Fig. 1. Architecture of XploRe, combining Client-Server and Layered architecture.

The presentation layer is responsible for displaying the application to the user and enabling interaction. It does not handle business logic. This is done via a web application accessible through any compatible browser using WebXR technology, which supports various display modes (Web, AR, and VR).

The business logic layer processes data input and handles requests, performing necessary operations. The controller component organizes the interaction between the view and model layers. Specific controllers manage user profiles, experiments, and GUI interactions. The model represents business logic elements, such as user profiles, experiments, and tests, while the GUI provides the interface for users to interact with the system.

The data layer is responsible for data persistence and storage. XploRe uses Firebase, a non-relational database from Google, where user data and interactions are stored in collections. Each user's session and usage data are stored in the "users" collection, which contains specific information on each student.

4   System Interfaces

The graphical user interface is one of the most important aspects of application development, as it is the part of the software with which the end user interacts. The interfaces of
XploRe were designed to be responsive and accessible to the target audience (secondary
school students). The interfaces are shared across different viewing modes, with the
primary difference being the way users interact with them. The interfaces of the XploRe
tool are presented with a brief description (see Fig. 2.).

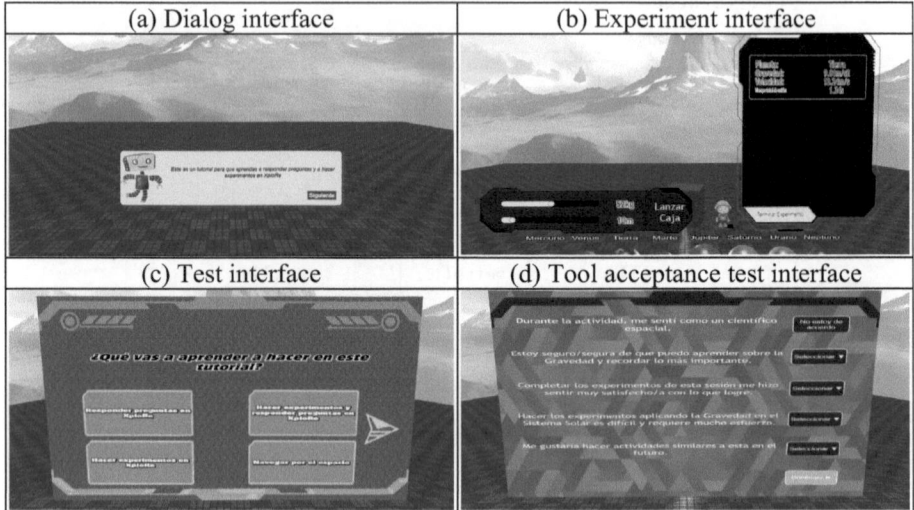

Fig. 2. Main interfaces of XploRe.

The dialog interface (a) includes a dialogue box featuring a robot drawing to capture
the student's attention, an instruction, and a 'next' button that allows the student to
proceed to the next dialogue or move to the next section if no further dialogues are
available.

In the experiment interface (b), the bottom left of the interface contains two yellow
sliders to adjust the object's weight (upper slider) and the fall distance (lower slider),
along with a button to launch the object. On the right, there is a notes section. If the
student requests help, steps for the activity appear; otherwise, the notes section remains
empty. It also includes a button to end the experiment. During the experiment, students
can interact with planets displayed at the bottom of the screen as part of the immersive
environment. To interact, students select parameters and launch the object, with the
system logging the experiment values in the notes section.

The test interface (c), which is used for comprehension questions, Pre-Test, and Post-
Test includes a background pattern, the question at the top, four answer buttons, and an
arrow button to review the question and proceed to the next one. For comprehension
questions, if the student answers incorrectly, pressing the continue button will return
them to the introductory dialogue to read it again.

The tool acceptance test (d) includes five questions per level with five Likert scale options: "Strongly disagree", "Slightly agree", "Somewhat agree", "Strongly agree", and "Completely agree". When a student selects a button next to each question, only the current question and its response options are displayed. After all the questions are answered, the "continue" button is enabled for the student to proceed.

5   Evaluation

To evaluate the effectiveness of the learning environment among second-grade secondary school students, tests were conducted using the tool, and the students' performance was assessed in each of the experiments within XploRe. The evaluation was performed in the middle school Simón Bolivar at Navolato, Sinaloa; the students were divided in two groups: experimental with 23 students (15 male and 8 female) and control with 23 students (14 male, 8 female and 1 did not specify); the age of the students was between 12 and 15 years old. To divide the students into the groups, an analysis of the general average of the grades was used, making both groups equally divided. Figure 3 shows the steps students follow for the evaluation.

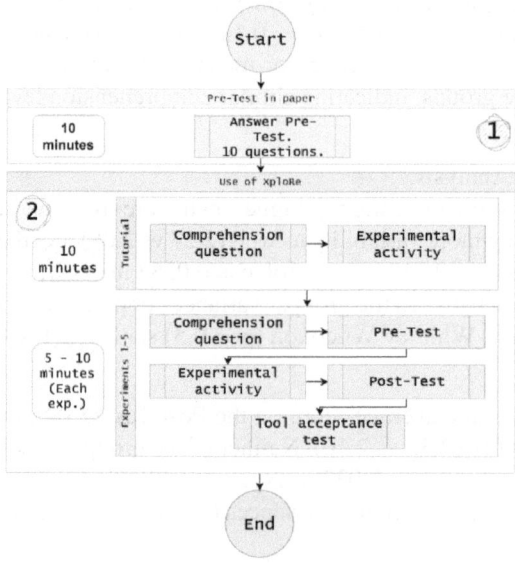

Fig. 3. Process of evaluation of XploRe.

The evaluation consisted of a paper-based Pre-Test that included topics regarding to the concept of gravity, acceleration, weight, and the law of gravity to determine the students' prior knowledge, followed by interaction with the five experiments within XploRe. In addition to the experiments, XploRe includes a tutorial to help students familiarize themselves with the tool. Each experiment contains a comprehension question, a specific Pre-Test and Post-Test for each experiment, and an experimental activity with which the student interacts. All the activities were designed based on the topics included in the second-grade student curriculum provided by Secretaría de Educación Pública (SEP).

6 Results

After the students finished the activities, all the data was collected from the database to be analyzed. The statistical analysis of the student questionnaires revealed several key findings:

Pre-Test Paper Analysis:
The control group had a slightly higher average score (5.83 vs 5.52), but the experimental group showed greater variability in results (1.97 vs 1.87). Both groups followed a normal distribution ($p = 0.263$ in control, $p = 0.567$ in experimental), and no significant differences were found in the variances ($F = 0.009$) or the mean scores ($p = 0.29$ one-tail, $p = 0.59$ two-tail) between the groups.

Comprehension Question Analysis:
The experimental group performed better on the comprehension question (2.69 vs 2.43), though they exhibited more variability (0.58 vs 0.75). The results showed a normal distribution ($p = 0.3785$ in control, $p = 0.1481$ in experimental) and no significant differences in variance ($F = 7.93E\text{-}15$) or mean scores ($p = 0.09$ one-tail, $p = 0.19$ two-tail) between the groups, indicating similar comprehension levels despite the slight differences in performance.

Pre-Test in the Tool Analysis:
The experimental group scored slightly higher on the Pre-Test conducted within the tool (5.48 vs 5.05) showing more variability as well (2.27 vs 2.08), but the data did not follow a normal distribution ($p = 0.0195$ in control, $p = 0.0383$ in experimental). There were no significant differences between the groups' variances or performance based on ANOVA ($F = 0.497$) and Mann-Whitney Wilcoxon ($p = 0.5$) statistical tests.

Post-Test Analysis:
Both groups had the same average score on the Post-Test (5.04), but the control group showed more variability (2.33 vs 2.16). Neither of the groups presented a normal distribution ($p < 0.01$ in control, $p = 0.033$ in experimental). No significant differences were found between the groups in terms of variances ($F = 0.026$) or performances ($p = 0.70$).

Correlation Analysis
The correlations between the various tests were weak or nearly null, indicating that performance in one test did not significantly impact performance in others. This suggests that the results were largely influenced by random variation rather than any specific patterns in student responses. These correlations can be observed in Table 1.

Table 1. Correlations of the tests in XploRe

	Pre-Test paper	Comprehension	Pre-Test tool	Post-Test
Pre-Test paper	1.000	−0.218	0.135	−0.098
Comprehension	−0.218	1.000	0.280	−0.021
Pre-Test tool	0.135	0.280	1.000	−0.184
Post-Test	−0.098	−0.021	−0.184	1.000

7 Discussion and Conclusions

The design and evaluation of XploRe faced several significant challenges throughout the development process, which were reflected in the results of student evaluations. Conducting the tests in a public school introduced limitations, including a lack of compatible equipment, which required testing in small groups instead of with the entire class. Additionally, the high temperature in the classroom due to a malfunctioning air conditioner made students uncomfortable, and an unstable internet connection caused delays and restarts for some students.

Another issue was the length of the experiments, which affected student motivation. By the time students reached the final exercises, their willingness to participate had decreased. Some students also reported feeling overwhelmed by the number of acceptance questions for the tool. These factors contributed to the low scores on the knowledge tests, with averages close to 5 out of 10, a failing grade in Mexico. For the comprehension question, students averaged nearly three attempts to answer correctly, indicating that most had to try almost all available options. The test scores were similar between both groups, showing no significant differences, and the Pre-Test and Post-Test data within the tool showed a non-normal distribution, suggesting that students may not have read the instructions carefully or engaged fully with the activities.

The research presented in this paper and the development of XploRe resulted in the achievement of the established objectives, as the tool was successfully developed to include activities that allow experimentation with the physical phenomena of the Solar System, an XR environment, and experiments that follow Bloom's Taxonomy. Additionally, XploRe enables the recording of students' responses and data to determine whether their learning improves with the use of XploRe. By analyzing other learning tools that share characteristics with XploRe, we can conclude that this research has led to the development of a tool that allows students to experiment in a more immersive and interactive manner, providing broad compatibility for PCs and mobile devices through the use of WebXR and the Web, AR, and XR viewing modes. The test results showed that students were not fully engaged in the activity, leading to very low average scores on all tests and a non-normal distribution of data in both the Pre-Test and Post-Test within XploRe. Various unforeseen issues during the testing process prevented optimal performance. The conclusion drawn is that further testing is necessary to gather statistically significant data.

For future work, evaluations with students must be redesigned to consider the factors that impacted the evaluation results: a smaller number of experiments should be chosen so that students do not find the number of activities and tests overwhelming. After the evaluations are conducted again, a statistical analysis should be performed to determine whether the hypotheses can be rejected or accepted with statistically significant data.

References

1. Sydon, T., Phuntsho, S.: Highlighting the importance of STEM education in early childhood through play-based learning: a literature review. RABSEL **22**(1) (2021)
2. Alves, P., Miranda, L., Morais, C.: The influence of virtual learning environments in students' performance. Univ. J. Educ. Res. **5**(3), 517–527 (2017)
3. Freina, L., Michella, O.: A literature review on immersive virtual reality in education: state of the art and perspectives. In: The International Scientific Conference E-learning and Software for Education, p. 8. Institute for Educational Technology, CNR, Genova (2015)
4. Huitt, W.: Bloom et al.'s taxonomy of the cognitive domain. Educ. Psychol. Interact. **22**, 1–4 (2011)
5. Torres-Samperio, G.A., de Jesús Gutiérrez-Sánchez, M., Suárez-Navarrete, A., Sánchez, D.H., Anaya, A.C.: Realidad extendida gamificada en la enseñanza de las ciencias naturales. Pädi Boletín Científico Ciencias Básicas Ingenierías ICBI **10**(Especial3), 69–79 (2022)
6. Sanson, B.: SpaceVR: outreach out of this world!. In: Wellington Faculty of Engineering Symposium (2023)
7. Rejón, L.A.G., Martínez, U.E.F., Galván, I.A., Rodríguez, E.B., Cano, G.Y.V.: Aplicación con realidad aumentada para apoyar al aprendizaje del sistema solar para alumnos de 5° y 6° de primaria en Tizayuca Hidalgo, en el periodo de 2020 al 2021. Bol. Científico INVESTIGIUM Escuela Superior Tizayuca **8**(16), 1–6 (2023)
8. Tresnawati, D., Fatimah, D.D.S., Rayahu, S.: The introduction of solar system using augmented reality technology. In: Journal of Physics: Conference Series, vol. 1402, no. 7, p. 077003. IOP Publishing (2019)
9. Huang, H.: Exploring the cosmos: prototype of a gesture-controlled vr system for interactive astronomical education. Doctoral dissertation, CALIFORNIA STATE UNIVERSITY SAN MARCOS (2024)
10. Ali, S., Alauldeen, R., Ruaa, A.: What is client-server system: architecture, issues and challenge of client-server system. HBRP Publ. **2**(1), 1–6 (2020)
11. Deacon, J.: Model-view-controller (MVC) architecture, vol. 28, p. 61 (2009). http://www.jdl.co.uk/briefings/MVC.pdf. Accessed 10 Mar 2006

Emotion Recognition in Virtual Reality Learning Environments: A Multimodal Machine Learning Approach

Luis Romero-Ramos$^{(\boxtimes)}$ ⓘ, Gabriel González-Serna ⓘ, Máximo López-Sánchez ⓘ, Nimrod González-Franco ⓘ, and Blanca Valenzuela-Robles ⓘ

Computer Science Department TecNM/CENIDET, 62490 Cuernavaca, Morelos, México
{m23ce080,gabriel.gs,maximo.ls,nimrod.gf,
blanca.vr}@cenidet.tecnm.mx

Abstract. Positive emotions significantly impact learning outcomes as they foster problem-solving and decision-making, while negative emotions can hinder information processing. Traditional methods for detecting emotions in immersive virtual reality (VR) applications, such as electroencephalography (EEG) and electrocardiography (ECG), are often invasive, expensive, and susceptible to noise. This paper introduces a new, non-invasive approach for classifying emotional valence in Virtual Immersive Learning Environments (VILE) based on analyzing head and hand movements using machine learning algorithms. A modified Self-Assessment Mannequin collected multimodal behavioral data from 30 students. The results show promising accuracy (71%) in distinguishing between positive and negative emotional states. While this method currently focuses on valence classification rather than specific emotion recognition, it paves the way for developing more accessible and less intrusive solutions for emotional detection in VR learning contexts.

Keywords: Virtual Reality · AVAI · SAM · Behavioral Analysis · Affective Computing · HCI

1 Introduction

Immersive Virtual Reality (IVR) has transformed education and training in the last decade. It offers highly engaging environments for exploration and learning [1]. IVR can simulate historical scenarios and microscopic worlds, providing levels of immersion that traditional methods cannot match. Its applications go beyond education, addressing ethical concerns in areas such as medicine through simulated surgical procedures [2]. Recognizing the potential of IVR for safe and cost-effective learning, various industries, including military, aviation, and manufacturing, have embraced this technology.

Virtual Immersive Learning Environments (VILE) have gained significant attention due to their ability to adapt learning experiences based on user emotions [1]. Affective Tutoring Systems (ATS), designed to personalize instruction according to emotional

L. Martínez-Villaseñor et al. (Eds.): MICAI 2024 Workshops, LNAI 15465, pp. 45–52, 2025.
https://doi.org/10.1007/978-3-031-83882-8_5

states, offer promising avenues for enhancing VILE [3]. Emotional valence, a measure of affective response as positive or negative, is a crucial component in emotion detection. Initially introduced by Russell (1980), emotional valence provides a simplified yet effective framework for classifying emotions, facilitating integration into affective interfaces that prioritize user personalization.

The traditional methods for detecting emotions often rely on physiological data, such as electroencephalography (EEG) and electrocardiography (ECG). However, these methods can be invasive, expensive, and susceptible to noise [4, 5]. To address these limitations, this paper proposes a new approach that uses multimodal behavioral data collected directly from the IVR viewer's sensors, specifically head and hand movements, to classify emotional valence.

This study aimed to evaluate the effectiveness of machine learning algorithms in classifying users' emotional valence while interacting with 360-degree videos in immersive virtual reality (IVR) environments. We analyzed behavioral data, specifically head and hand movements, to automatically classify emotional valence and compare these results to user-reported valence ratings.

The following hypotheses were tested:

- H0: There is no significant difference between the emotional valence classifications made by machine learning algorithms based on behavioral data and those reported by users when viewing 360-degree videos in IVR environments.
- H1: A significant difference exists between the emotional valence classifications made by machine learning algorithms based on behavioral data and those reported by users when viewing 360-degree videos in IVR environments.

This research can potentially enhance VILE and STA systems by enabling more precise and less invasive personalization of learning experiences. By adapting to user emotions without relying on expensive or intrusive physiological sensors, this approach could revolutionize adaptive learning environments by implementing emotional interfaces. While this study focused on emotional valence classification, it represents a promising step toward developing more sophisticated and effective adaptive learning systems.

2 Methodology

2.1 Study Design

An experimental study evaluated supervised machine learning algorithms' capability to classify emotional valence using behavioral data collected in an immersive virtual reality (IVR) environment. The study involved 30 participants split into two groups of 15. Each group watched four 360-degree videos, and head and hand movement data were continuously recorded using a Meta Quest 3 device.

The Meta Quest 3's tracking system, Oculus Insight, uses a combination of inertial measurement units (IMUs), RGB cameras, a depth sensor, and infrared cameras to accurately track the headset and controllers' position and orientation at a rate of 60 Hz with an average error of 1.1 cm [6, 7]. This reliable tracking system ensures accurate data collection for analysis.

After watching each video, the participants used an adapted version of the Self-Assessment Manikin (SAM) instrument [8] to assess their perceived emotional valence. These ratings were used as reference labels to train the classification algorithms.

2.2 Participants

The study involved 30 graduate-level students (N = 30), comprising 22 males and eight females, with an average age of 28 years. Participants were randomly divided into two groups, each consisting of 15 individuals.

2.3 Stimuli

Two sets of four 360-degree videos were chosen for the study. Each set included two videos with positive emotional content and two with harmful emotional content. The videos, which were about 2 min long on average, were obtained from two primary sources:

- Public Repository: Half of the videos were taken from a public repository. These videos were initially used in a previous study by [9] and had already been evaluated for emotional content and intensity.
- Online Sources: Based on the recommendations of the same study, four more videos were selected from online platforms in addition to those from the public repository.

2.4 Procedure

The research was conducted in a controlled laboratory environment where participants sat in swivel chairs. Before the experiment, they received detailed instructions on using the Meta Quest 3 immersive virtual reality (IVR) device and provided their gender and age via an in-app form.

Participants viewed four consecutive 360-degree videos, during which the Meta Quest 3's tracking system recorded their head and hand movements—capturing rotation (yaw, pitch, roll) and translation across the X, Y, and Z axes at 60 Hz.

After each video, participants rated their emotional valence using an adapted version of the Self-Assessment Mannequin (SAM) instrument [8], classifying their responses as positive or negative to serve as ground truth for training machine learning algorithms. The collected data included various head and hand movement variables, detailed in Table 1.

2.5 Data Pre-processing

We collected 120 data files from the experiment (four per participant), but 13 were excluded due to user interruptions or low battery levels in the Meta Quest 3 device, leaving 107 usable files. Each file contained approximately 9000 head and hand position records and rotation data records. Min-max normalization was applied to prepare the data for analysis using Python and the Pandas library, ensuring that all values fell within the range of 0 to 1. The processed data were saved in CSV format.

Table 1. Head and hand movement variables

No.	Variable	Description
1	Head_Position	Head position on the x, y, and z axes
2	Head_Rotation	Head rotation in the x, y, and z axes
3	Left_Hand_Position	Position of the left hand on the x, y, and z axes
4	Left_Hand_Rotation	Left-hand rotation on the x, y, and z axes
5	Right_Hand_Position	Position of the right hand in the x, y, and z axes
6	Right_Hand_Rotation	Right-hand rotation on the x, y, z axes
7	Valence	Valence level (0 to 10)

As noted in previous research [4, 9], there is a potential correlation between emotional valence and the standard deviation of head turns. We analyzed the recorded head rotation angles (pitch, yaw, and roll) over time to explore this. The results, shown in Fig. 1, indicate that while pitch and roll movements were subtle, yaw, representing side-to-side head turns, exhibited more variation, supporting the hypothesis of a link between head movements and emotional valence.

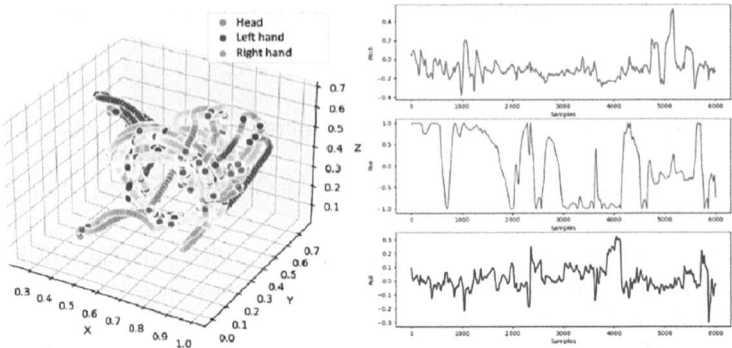

Fig. 1. 3D head, right, and left-hand positions (left). Varying head rotation angles (right)

2.6   Statistical Data Processing

A comprehensive statistical analysis of the individual data files was conducted, including calculations of mean, standard deviation, maximums, and minimums for head and hand positions and rotations. This analysis provided valuable insights into variability and behavioral patterns during video viewing. Additionally, composite measurements, such as hand-to-head distances, were calculated. Each data file contained approximately 9,000 records subjected to statistical analysis to derive 32 feature vectors. These vectors included descriptive statistics and composite measures. Each feature vector was associated with an emotional valence label (1 for positive, 0 for negative) based on participant responses to the adapted SAM instrument administered after each video.

After processing, 102 feature vectors were generated to form the training dataset. Five records were discarded due to inconsistencies caused by missing data, likely resulting from user errors or device malfunctions.

3   Results

3.1   Model Performance

Three supervised machine learning algorithms were evaluated for their ability to classify emotional valence from the collected data:

- Support Vector Machine (SVM): Known for its effectiveness in class separation.
- Random Forest: Renowned for its robustness to noise and its ability to assess feature importance.
- Multilayer Perceptron Neural Network (MLP): Capable of modeling complex nonlinear relationships.
- These algorithms were selected for their suitability for handling nonlinear and high-dimensional data.

The dataset comprised 102 instances with 32 features and two negative and positive valence labels. To train and test the models, the data were randomly divided into 70% for training and 30% for testing. Figure 2 illustrates the distribution of classes within the dataset.

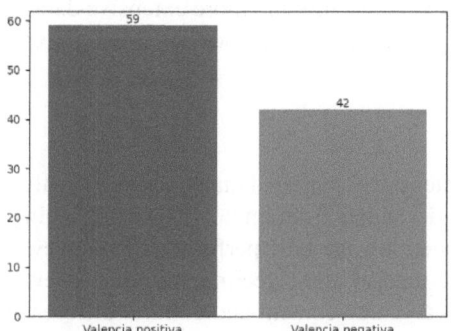

Fig. 2. Distribution of dataset classes

The performance of each model was assessed using the Accuracy, Recall, and F1-score metrics, and the results are presented in Table 2.

Table 2. Evaluation Metrics

Algorithm	Precision	Recall	F1-Score
SVM	0.64	0.63	0.61
Random Forest	0.61	0.6	0.59
MLP	**0.71**	**0.7**	**0.7**

3.2 Evaluation of Results with McNemar Test

A McNemar test was conducted to evaluate the performance of machine learning algorithms by comparing their classifications to self-reported valences. The critical McNemar value at a 0.05 significance level was 3.841.

The results were as follows:

- Support Vector Machine (SVM): The test statistic was approximately 3, below the critical value, indicating no significant difference between SVM classifications and self-reported valences, leading to acceptance of the null hypothesis (H0).
- Random Forest: The test statistic was approximately 1.333, also below the critical value, maintaining the null hypothesis and suggesting no significant difference from self-reported valences.
- Multilayer Perceptron Neural Network (MLP): The test statistic was 0.1111, below the critical value, providing insufficient evidence to reject the null hypothesis.

While all algorithms effectively classified emotional valence, the McNemar test showed no significant differences between their classifications and self-reported valences. This suggests the models capture overall emotional response trends but highlights the need for improvements through more extensive datasets and the integration of additional sensory technologies, like eye tracking or facial expression analysis.

4   Discussion

The research demonstrates the potential of machine learning algorithms to classify affective valence using behavioral data from immersive virtual reality environments, with the MLP neural network showing the best performance. However, improvements can be made by expanding the dataset and selecting stimuli that evoke consistent emotional responses. Factors such as video content and cultural differences may affect emotional variability, highlighting the need for diverse samples to enhance generalizability.

Limitations include a small, homogeneous sample size, which may restrict the findings' applicability, and the subjectivity of self-assessment could introduce biases. Additionally, the short duration of experimental sessions and controlled conditions should be considered in interpretation.

Future research should assess the real-time efficiency of machine learning algorithms in Virtual Immersive Learning Environments or Affective Tutoring Systems. The Sentis library [10] allows the integration of neural network models from TensorFlow or PyTorch into Unity for use on Meta Quest 3. Incorporating sensors like eye tracking and facial expression analysis could improve the accuracy of affective valence classification and specific emotional identification, which is crucial for enhancing learning outcomes in these environments.

5 Conclusions

This study demonstrates the potential of machine learning algorithms for classifying emotional valence in immersive virtual reality environments using behavioral data. Among the evaluated models, the Multilayer Perceptron Neural Network achieved the highest accuracy of 71%. However, statistical tests indicated no significant differences between the algorithm-generated classifications and self-reported valences.

While these results are encouraging, the inherent complexity of accurately detecting emotions from behavioral data remains a significant challenge. Emotional valence is just one aspect of the emotional spectrum, and capturing more complex emotional states necessitates more sophisticated approaches. Furthermore, incorporating additional data sources and features, such as eye tracking or facial expression analysis, could enhance the performance of current models.

Although the evaluated algorithms show promise for applications in Virtual Immersive Learning Environments and Affective Tutoring Systems, further research is needed to address the complexities of emotional detection. Continued advancements in machine learning models and the expansion of feature sets will be crucial for developing more accurate and robust emotional detection systems.

References

1. Marín-Morales, J., Llinares, C., Guixeres, J., Alcañiz, M.: Emotion recognition in immersive virtual reality: from statistics to affective computing. Sensors (Basel) **20**, 5163 (2020). https://doi.org/10.3390/s20185163
2. Xie, B., et al.: A review on virtual reality skill training applications. Front. Virtual Reality **2** (2021)
3. Hasan, M.A., Noor, N.F.M., Rahman, S.S.B.A., Rahman, M.M.: The transition from intelligent to affective tutoring system: a review and open issues. IEEE Access **8**, 204612–204638 (2020). https://doi.org/10.1109/ACCESS.2020.3036990
4. Holzwarth, V., et al.: Towards estimating affective states in Virtual Reality based on behavioral data. Virtual Reality **25**, 1139–1152 (2021). https://doi.org/10.1007/s10055-021-00518-1
5. Velázquez-Cano, J.E., Gonzáles-Serna, J.G., Rivera-Ri, L.: Method to identify emotions in immersive virtual learning environments using head and hands spatial behavioral information (2023)
6. Abdlkarim, D., et al.: A Methodological Framework to Assess the Accuracy of Virtual Reality Hand-Tracking Systems: A case study with the Oculus Quest 2 (2022). https://doi.org/10.1101/2022.02.18.481001. https://www.biorxiv.org/content/10.1101/2022.02.18.481001v1
7. Powered by AI: Oculus Insight. https://ai.meta.com/blog/powered-by-ai-oculus-insight/. Accessed 23 Sept 2024
8. Betella, A., Verschure, P.F.M.J.: The affective slider: a digital self-assessment scale for measuring human emotions. PLoS ONE **11**, e0148037 (2016). https://doi.org/10.1371/journal.pone.0148037

 9. Li, B.J., Bailenson, J.N., Pines, A., Greenleaf, W.J., Williams, L.M.: A public database of immersive vr videos with corresponding ratings of arousal, valence, and correlations between head movements and self report measures. Front. Psychol. **8** (2017)
10. Sentis overview | Sentis | 2.1.0. https://docs.unity3d.com/Packages/com.unity.sentis@2.1/manual/index.html. Accessed 23 Sept 2024

Enhancing Dropout Prediction Models Through Feature Selection Techniques

Daniel Domínguez-Gómez[1] , Eduardo Sánchez-Jiménez[2]([⊠]) ,
Yasmín Hernández[2] , Juan de Dios González Torres[1] ,
and Javier Ortiz-Hernandez[2]

[1] Information Technology Department, UJAT-DACyTI, Cunduacan, Mexico
231H20007@alumno.ujat.mx, juan.gonzalez@ujat.mx
[2] Computer Science Department, TecNM/Cenidet, Cuernavaca, Mexico
{d22ce005,yasmin.hp,javier.oh}@cenidet.tecnm.mx

Abstract. School dropout poses a significant threat to personal and social development, limiting employment opportunities and perpetuating economic inequality. This study addresses the challenge of identifying student dropouts by focusing on selecting relevant features or variables. We applied five feature selection techniques: ANOVA, mutual information, sequential forward selection, recursive feature elimination, and the least absolute shrinkage and selection operator, highlighting key sociodemographic features. Fifteen machine learning models, including Decision Trees, Support Vector Machines, and K-nearest neighbors, were trained using these techniques and evaluated through cross-validation. The results showed that the mutual information KNN model achieved the highest precision and F1 score, while Sequential Forward Selection provided optimal results for Decision Trees and Support Vector Machines. These results highlight the critical role of feature selection techniques in identifying the variables that significantly affect the effectiveness of predictive models.

Keywords: dropout prediction · education · feature selection techniques · machine learning

1 Introduction

Education is essential for individual and societal advancement, providing knowledge and opportunities that shape personal lives and drive community progress, leading to a more prosperous society. However, the phenomenon of school dropout represents a significant challenge to this process. According to Kuz and Morales [1], school dropout impedes student intellectual growth, restricting their access to desirable employment opportunities and, consequently, reducing their quality of life. This problem is especially concerning at the university level, where student dropout has emerged as a significant challenge for educational institutions in recent years. The loss of students at any academic level represents a

L. Martínez-Villaseñor et al. (Eds.): MICAI 2024 Workshops, LNAI 15465, pp. 53–60, 2025.
https://doi.org/10.1007/978-3-031-83882-8_6

substantial loss of human capital, but in the context of higher education, the repercussions are particularly severe.

Machine learning has been successfully applied in various areas, such as identifying at-risk students, predicting final exam grades, and early detection of students at risk of failing (Pek et al., 2021). Additionally, feature selection (FS) is an essential step in the modeling process. It helps identify the most relevant features while eliminating irrelevant or redundant ones, reducing model complexity, and improving predictive accuracy and reliability.

2   Background

2.1   Education and Dropout

Tinto [2] conceptualizes dropout as the situation experienced by a student who aspires to complete his educational project but fails to do so. He points out that a dropout can be classified as an individual who, as a student, does not participate in academic activities for three consecutive semesters. González [3] proposes a more detailed classification of dropout at the university level: initial, early, or late, depending on when the student abandons his or her studies. In terms of the spatial factor, there are three main categories of dropout: institutional, internal, and dropout from the education system. Institutional dropout occurs when the dropout occurs in a specific institution. Internal dropout refers to a situation where the student moves within the same university. Dropout from the education system is when the student leaves higher education altogether.

2.2   Machine Learning and Feature Selection Techniques

The models were built using three widely used machine learning algorithms: Decision Trees (DT), Support Vector Machines (SVM), and K-Nearest Neighbors (KNN). These algorithms were trained on five datasets, each generated using a different feature selection technique.

DT is a classification method does not store the training data in its entirety. Instead, it uses the training data to build a stratification or tree structure that recursively partitions the training space into regions with similar labels or features. One of its key features is recursive partitioning, which allows the dependent variable to be continuous, discrete, or categorical, and the predictor variables to be continuous, discrete, or categorical [4].

SVM is based on the principles of statistical learning theory. The goal is to create a clear separation between different categories of data by identifying the optimal hyperplane that best separates the classes and maximizes the margin between the two classes. This approach facilitates superior generalization and prediction of previously unseen data [5].

KNN is a method based on instance-based learning. The process involves making comparisons to assign a new instance to the majority class of its k nearest neighbors. This algorithm is straightforward yet effective for classification tasks. It is crucial to select an appropriate value of k to ensure optimal results [6].

The application of machine learning algorithms to data sets comprising a high number of features can result in a number of issues that impact the efficacy of the algorithms. These include overfitting, increased computational and learning time, and model degradation in the presence of noisy data. One of the most effective solutions to these issues is the use of dimensionality reduction techniques. This technique involves the selection of a subset of features that contain the most relevant information, as described in reference [7].

According to Li et al. [8], working with an optimized dataset of significant features can significantly improve the predictive performance of a machine learning model, reduce model complexity, lower computational and resource costs, and prevent algorithm overfitting. Although domain experts can identify and eliminate some irrelevant attributes, the optimal selection of a subset of features generally requires a more structured and systematic approach. In the following, we discuss the techniques we apply to the dropout dataset.

Analysis of Variance (ANOVA) is one of the most widely used statistical techniques with applications across the full range of experiments in agriculture, biology, chemistry, toxicology, pharmaceutical research, clinical development, psychology, social sciences, and engineering; the procedure involves separating the total observed variation in the data into individual components attributable to different factors as well as those caused by random or chance variation [9].

Mutual Information (MI) measures the information shared between features and the target variable [10]. It is independent of any machine learning algorithm and is based on results from various statistical tests.

Sequential Forward Selection (SFS) adds features one at a time, selecting the one that improves the model the most at each step. Recursive Feature Extraction (RFE): Selects features by iteratively eliminating those that have the least impact on model performance. These methods use a subset of features to build a model and decide to add or remove features from that subset based on model performance [11].

The Least Absolute Shrinkage and Selection Operator (LASSO) performs a regularization that forces some feature coefficients to be exactly zero, thereby removing them from the model. These methods select features during the execution of the learning algorithm and are integrated as part of the learning process [12].

2.3   Related Work

This section shows the existing works on predictive models for student dropout. Examines both approaches that did not apply machine learning techniques and those that did. In addition, it is the various feature selection methods commonly used in these studies.

Talamas and Ceballos [13] propose an ensemble stacking technique as a means of combining predictive models that, with a limited number of variables, can achieve the desired results for the prediction of early dropout. The results demonstrate that implementing this technique in an intervention program would be a cost-effective strategy. In contrast, Cuevas-Chávez et al. [14] focused on parameter optimization and resampling techniques (SMOTE, ADASYN, SMOTE-SVM, and SMOTE + ENN) to address data imbalance. They evaluated the performance of classifiers such as Random Forest (RF), Support Vector Machine (SVM), and XGBoost. The combination of SMOTE+ENN with the SVM classifier demonstrated the most optimal performance, with an accuracy rate of 94.11%. Both studies primarily focused on algorithm and hyperparameter tuning, with minimal attention devoted to feature selection techniques.

Works on feature selection for school dropout prediction models show significant improvements in machine learning performance. Youssef et al. [15] applied SVM, Decision Tree, Naive Bayes, LR, and KNN to a dataset with 5,327 records using techniques like Sequential Forward Selection and Recursive Feature Elimination. After applying SVM, their model improved to an AUC of 0.97, precision of 0.98, recall of 0.98, and F1-score of 0.98. Zapata-Medina et al. [16] used a dataset of 1,865 records and 29 features. After feature selection with Random Forest, precision reached 0.92, recall 0.57, and F1-score 0.70. Lin Qui et al. [17], working with a large MOOC dataset, also applied feature selection methods like Mutual Information and RFE, achieving stable results, with an AUC of 0.86 and precision of 0.85 in their Logistic Regression model.

3 Dropout Prediction

3.1 Dataset

This study utilizes the dataset *Predict Student Dropout and Academic Success*, which is available in the UCI Machine Learning Repository[1]. The dataset comprises 4,424 instances with 34 features, including 20 discrete, 8 binary, 5 continuous, and 1 ordinal feature. These features encompass demographic, macroeconomic, socioeconomic, and enrollment information for students across 17 university programs. The target variable is classified as follows: Dropout, Graduate, and Enrolled.

3.2 Exploratory Data Analysis

A correlation analysis was conducted to identify strongly correlated independent features to improve feature selection and avoid multicollinearity. Heatmaps were employed to quantify the correlations, which revealed a high correlation between variables such as *Nationality* and *International, parents' qualifications* and *professions*, and *curriculum units in semesters 1 and 2*. Specifically, strong correlations were found between various curricular units from both semesters, as detailed in Table 1.

---

[1] Predict students' dropout and academic success: https://doi.org/10.24432/C5MC89.

Table 1. Correlation between curricular units in 1st and 2nd semesters.

Curricular units 1st semester	Curricular units 2nd semester
Enrolled	Credited
Credited	Enrolled
Evaluations	Evaluations
Approved	Approved
Grade	Grade
Without evaluations	Without evaluations

3.3 Preprocessing

Records labeled as *Enrolled* were excluded from analysis, preprocessing, and model construction as they were not useful for our research. This resulted in 3,630 records for the *Dropout* and *Graduate* classes, respectively. Furthermore, the features *Nationality, mother's occupation, and father's qualification* were excluded from the analysis due to their collinear and redundant nature.

In contrast, the features associated with the curricular units of the first and second semesters were subjected to a Principal Component Analysis (PCA) to reduce the dimensionality of the data set and mitigate the effects of multicollinearity between these variables without loss of information. Subsequently, the data were scaled using the StandardScaler method to ensure that they were all on the same scale, thus guaranteeing the correct interpretation and behavior of the models.

In the study by Cuevas-Chavez et al. (2024), they addressed the issue of class imbalance to improve the performance of the model. Among the techniques used, SMOTE+ENN yielded the best results. Based on this reference, we implemented SMOTE+ENN in our current work. Before its application, the dataset was unbalanced, with 2209 instances of graduates and 1421 instances of dropouts. After balancing the data, the number of dropout instances increased to 1657, while the number of graduate instances was adjusted to 1602. This suggests that SMOTE+ENN effectively balanced the classes by increasing the minority class instances (*dropouts*), thereby improving the model's ability to learn from these instances.

3.4 Feature Selection

The common features selected by the techniques are listed below. This pattern illustrates the importance of these features in predicting dropout, regardless of the feature selection method used.

- Marital status, application mode, daytime/evening attendance, father's occupation, debtor, tuition fees up to date, scholarship holder, age at enrollment, curricular 1st and 2nd sem PCA.

3.5   Hyperparameters Configuration

As part of the experimental setup for the training process with the five feature selection techniques, the hyperparameters of the algorithms were adjusted. For the decision tree, different values of max_depth (ranging from 1 to 14) were tested. The support vector machines were evaluated with kernels such as linear, poly, RBF, and sigmoid. In addition, the KNN model was trained using odd values for the number of neighbors, including 1, 3, 5, 7, 9, 11, 13, and 15.

4   Results and Discussion

In this study, 15 models were trained and evaluated using 10-fold cross-validation. These models are the result of combining three machine learning classifiers with five different feature selection techniques. In addition, we used four key metrics: accuracy, precision, recall, and F1 score to evaluate the trained models.

The Fig. 1 shows results precision metrics in test phase. The DT algorithm, when combined with the SFS technique, achieved the highest precision at 95.32%, while the accuracy, recall, and F1-score were each 94.30%. For the SVM algorithm, the SFS technique also delivered optimal results, with a precision of 94.21% and accuracy, recall, and F1-score all at 93.99%. In contrast, the KNN algorithm paired with the MI technique proved the most effective, attaining a precision of 99.05%, with accuracy, recall, and F1-score matching at 99.05%.

Fig. 1. Precision scores trained models

These findings suggest that the KNN model, using the MI technique, offers the best predictive performance, as it achieved the highest scores across all metrics. Additionally, the SFS technique demonstrated superior results for the DT and SVM algorithms, outperforming other feature selection methods.

Table 2 shows the confusion matrix and the results of the testing models.

Table 2. Testing classifiers performance.

Algorithm	FS Technique	TN	FN	FP	TP	P_r	A_c	R_e	F_1
DT	ANOVA	359	15	15	374	94.37	94.33	94.33	94.33
	MI	361	20	12	397	95.30	95.28	95.28	95.28
	SFS	358	22	13	372	95.32	94.30	94.30	94.30
	RFE	359	34	22	387	93.57	94.48	94.48	94.48
	LASSO	359	34	22	387	93.98	94.48	94.48	94.48
SVM	ANOVA	361	12	26	363	93.98	93.83	93.83	93.83
	MI	363	18	33	376	93.77	93.83	93.83	93.83
	SFS	366	14	25	360	94.21	93.99	93.99	93.99
	RFE	369	24	35	366	94.11	93.98	93.98	93.98
	LASSO	377	20	28	381	93.94	93.82	93.82	93.82
KNN	ANOVA	370	3	3	386	98.76	98.75	98.75	98.75
	MI	378	3	4	405	**99.05**	**99.05**	**99.05**	**99.05**
	SFS	373	7	4	381	98.83	98.82	98.82	98.82
	RFE	384	9	4	397	98.62	98.61	98.61	98.61
	LASSO	379	1	14	408	98.74	98.72	98.72	98.72

5   Conclusions and Future Work

In this study, five feature selection techniques were applied to a dropout prediction dataset, successfully reducing the feature space for machine learning models. The results indicated that the KNN algorithm, combined with the MI technique, achieved the highest performance across all metrics, consistently surpassing other models. Additionally, the SFS technique was particularly effective for optimizing the DT and SVM models. Importantly, all feature selection methods identified socio-demographic factors, such as marital status, father's occupation, and age at enrollment, as key predictors of dropout. These variables play a critical role in influencing academic outcomes and should be prioritized in future dropout prediction models.

In future work, incorporating advanced classification algorithms like neural networks could further enhance model performance. Moreover, exploring bio-inspired algorithms to discover additional relevant features may improve both predictive accuracy and insights into the factors driving dropout, ultimately leading to more targeted and effective educational interventions.

References

1. Kuz, A., Morales, R.: Ciencia de datos educativos y aprendizaje automático: Un caso de estudio sobre la deserción estudiantil universitaria en méxico. Educ. Knowl. Soc. (EKS), vol. e30080 (2023)
2. Tinto, V.: Limits of theory and practice in student attrition. J. High. Educ. **53**(6), 687–700 (1982)
3. González, L.E.: Estudio sobre la repitencia y deserción en la educación superior chilena. Digital Observatory for Higher Education in Latin America and The Caribbean, IESALC-UNESCO (2005)
4. Ramasubramanian, K., Singh, A.: Machine Learning Using R, no. 1. Springer, Cham (2017)
5. Cristianini, N., Shawe-Taylor, J.: An Introduction to Support Vector Machines: And Other Kernel-Based Learning Methods. Cambridge University Press, Cambridge (1999)
6. Coomans, D., Massart, D.L.: Alternative k-nearest neighbour rules in supervised pattern recognition: part 1. k-nearest neighbour classification by using alternative voting rules. Anal. Chim. Acta **136**, 15–27 (1982)
7. Alonso-Betanzos, A.: Filter methods for feature selection. A comparative study. In: Proceedings of the International Conference on Intelligent Data Engineering and Automated Learning, UK, Birmingham, pp. 178–187 (2007)
8. Li, J., et al.: Feature selection: a data perspective. ACM Comput. Surv. **50** (2018)
9. Kaufmann, J., Schering, A.: Analysis of variance anova. Wiley Encyclopedia of Clinical Trials (2007)
10. Talavera, L.: An evaluation of filter and wrapper methods for feature selection in categorical clustering. In: Proceedings of the International Symposium on Intelligent Data Analysis, Spain, Madrid, pp. 440–451 (2005)
11. Karegowda, A., Manjunath, A., Jayaram, M.: Feature subset selection problem using wrapper approach in supervised learning. Int. J. Comput. Appl. **1**, 13–17 (2010)
12. Jović, A., Brkić, K., Bogunović, N.: A review of feature selection methods with applications. In: Proceedings of the 38th International Convention on Information and Communication Technology, Electronics and Microelectronics, Croatia, Opatija, pp. 1200–1205 (2015)
13. Talamás-Carvajal, J., Ceballos, H.: A stacking ensemble machine learning method for early identification of students at risk of dropout. Educ. Inf. Technol. **28**, 12169–12189 (2023)
14. Cuevas-Chávez, P., Narciso, S., Sánchez-Jiménez, E., Pérez, I., Hernández, Y., Ortiz-Hernandez, J.: School dropout prediction with class balancing and hyperparameter configuration. In: Advances in Computational Intelligence. MICAI. LNCS, vol. 14502, pp. 21–32. Springer, Cham (2023)
15. Youssef, M., Mohammed, S., Hamada, E., et al.: A predictive approach based on efficient feature selection and learning algorithms' competition: case of learners' dropout in MOOCs. Educ. Inf. Technol. **24**, 3591–3618 (2019)
16. Zapata-Medina, D., Espinosa-Bedoya, A., Jiménez-Builes, J.: Improving the automatic detection of dropout risk in middle and high school students: a comparative study of feature selection techniques. Mathematics **12**, 1776 (2024)
17. Qiu, L., Liu, Y., Liu, Y.: An integrated framework with feature selection for dropout prediction in massive open online courses. IEEE Access **6**, 71 474–71 484 (2018)

Assessing Cognitive Load
in Programming Exercises Based
on Readability and Lexical Richness

Jesús Miguel García-Gorrostieta[1]([✉]), Samuel González-López[2],
Aurelio López-López[3], Ulises Ponce-Mendoza[1],
and José David Madrid-Monteverde[1]

[1] Universidad de la Sierra, Moctezuma, Sonora, Mexico
{jgarcia,upmendoza,dmadrid}@unisierra.edu.mx
[2] Technological Institute of Nogales, Sonora, Mexico
samuel.gl@nogales.tecnm.mx
[3] Instituto Nacional de Astrofísica, Óptica y Electrónica,
Tonantzintla, Puebla, Mexico
allopez@inaoep.mx

Abstract. In this paper, we explore the impact of problem statement
readability and lexical richness on cognitive load during programming
exercises. Cognitive load theory suggests that the complexity of instructions can significantly affect student performance, especially in tasks
requiring high mental effort. Using NASA-TLX, this study measures students' cognitive load working on JavaScript and React programming problems with varying readability levels. Results show that students with
higher programming proficiency tend to experience lower cognitive load,
even with more complex problem descriptions, while less experienced students benefit from simplified instructions. This paper provides recommendations for instructional design to optimize cognitive load based on student proficiency, that could be incorporated in a learning environment.

Keywords: cognitive load · readability · lexical richness ·
programming training · programming exercises

1   Introduction

In programming training, the cognitive load imposed by exercise instructions and
the complexity of the required code are often overlooked. Both factors directly
affect students' ability to understand and solve problems in timely manner.
Therefore, it is crucial to consider the complexity of exercises relative to students' abilities. This paper explores the relationship between the cognitive load
experienced by students during programming tasks and the readability and lexical richness of the exercise instructions. It also proposes instructional design recommendations to minimize unnecessary cognitive load while optimizing learning
outcomes.

L. Martínez-Villaseñor et al. (Eds.): MICAI 2024 Workshops, LNAI 15465, pp. 61–70, 2025.
https://doi.org/10.1007/978-3-031-83882-8_7

We first present an overview of cognitive load theory and its implications in educational contexts. Next, we describe how cognitive load is measured, followed by the design of two programming exercises administered to students. Finally, we analyze the results and conclude how to best tailor instructional materials to reduce cognitive load and improve student performance.

2   Cognitive Load

Cognitive load is generally defined as the demands on working memory resources required to learn and solve a task or learning problem. Working memory is a reservoir of cognitive resources that are invested in learning tasks. This is used to temporarily retain and manipulate information for tasks performed in daily life. Working memory is temporary, only retaining information for a few seconds. It can only hold five to seven elements at a time, meaning it has limited capacity [1]. Working memory relies on attention control and mental effort [8].

Within cognitive load theory, there are three types of cognitive load: a) intrinsic, b) extrinsic, and c) relevant (or germane). Intrinsic load refers to the inherent cognitive demand of the task and the learner's level of expertise. Extrinsic cognitive load is related to unnecessary load that overwhelms, contaminates, and affects working memory [8]. Germane or relevant cognitive load occurs when working memory is focused on learning and the automation of schemas [6].

Programming involves handling complex concepts simultaneously, so cognitive load theory suggests reducing unnecessary load (extrinsic) and managing content complexity (intrinsic) to facilitate the construction of functional mental schemas (germane load) [2]. Some studies have incorporated physiological (such as heart rate and pupil dilation) and behavioral metrics to assess cognitive load. In [4] the authors analyzed how these measures can improve predictive models of cognitive load and customize educational systems to reduce mental overload in advanced programming tasks.

2.1   Measuring Cognitive Load

Several tools are used to measure cognitive load, such as NASA-TLX, FPSICO, ISTAS21, and MERS, which assess mental load, mental effort, and performance.

The NASA-TLX (Task Load Index) tool is based on the assumption that mental load is a hypothetical construct representing the cost incurred by an individual when trying to achieve a specific performance level. The level of mental load emerges from the interaction between task demands, the circumstances under which the task is performed, and the operator's skills, behaviors, and perceptions [5]. NASA-TLX distinguishes six dimensions of mental load, as follows:

Effort: The degree of mental and physical effort required by the subject to achieve their performance level.
Mental demand: The amount of mental and perceptual activity required by a task (e.g. thinking, deciding, calculating, remembering, looking, or searching).

Physical demand: The amount of physical activity required by a task (e.g. pressing, pushing, turning, and so on).

Temporal demand: The level of time pressure felt, i.e., the ratio between the time required and the time available.

Performance: The extent to which the individual feels satisfied with their level of performance.

Frustration level: The extent to which the subject feels insecure, stressed, irritated, dissatisfied, etc., during task execution [5].

2.2   Recommendation for the Construction of Educational Material

To facilitate the learning process is necessary to segment and sequence concepts. Another key aspect is to provide information in the clearest and most structured possible way. Additionally, the most direct and simple resources should be presented (e.g., using an image instead of a description). Below, several principles to consider when designing programming exercises are listed:

Open-Ended Problem Principle: Cognitive load is reduced when the problem does not have a single solution. A single-solution problem is one where all students must reach the same solution, while in an open-ended problem, each student arrives at their own solution [8].

Worked-Example Principle: This instructional application follows the principle of "borrowing," which means providing learners with examples of solved problems as an expert in the field would solve them [7]. This principle is utilized in the experimental exercises to reduce the cognitive load on students.

Problem-Completion Principle: Completion problems are partially solved problems, except that the learner must complete part of the answer [9].

Imagination Principle: This principle suggests asking learners to mentally review the procedures or concepts without the learning materials they have already worked with [8].

3   Readability and Lexical Richness

The readability and lexical richness are part of the structure of problem formulation in programming. Therefore, we consider the metrics of Flesch Reading Ease index and the lexical richness of the descriptions measured using indices such as the Type-Token Ratio. Below, we present the metrics used in the evaluation of programming descriptions.

Flesch Reading Ease. The Flesch Reading Ease formula, developed by Rudolph Flesch, is used to assess the difficulty level of English text. It provides an estimate of how easily a document can be understood, based on factors such as word length, sentence length, word form, and the number of syllables or letters. The formula generates a score, categorizing the readability level (Table 1 [3]).

Table 1. Flesch Reading Ease scores for assessing document readability.

Score	School Level	Description
100.0–90.0	5th Grade	Very easy to read. Easily understood by an average 11-year-old.
90.0–80.0	6th Grade	Easy to read. Conversational English.
80.0–70.0	7th Grade	Fairly easy to read.
70.0–60.0	8th-9th Grade	Plain English, easily understood by students.
60.0–50.0	10th-12th Grade	Fairly difficult to read.
50.0–30.0	College	Difficult to read.
30.0–10.0	College Graduate	Very difficult to read.
10.0–0.0	Professional	Extremely difficult to read, suitable for highly educated readers.

The formula is represented as:

$$\text{Score} = 206.835 - (1.015 \times \text{sw}) - (84.6 \times \text{wl})$$

where: wl = Word Length. sl = Average Sentence Length (the number of words divided by the number of sentences).

4   Cognitive Load in Programming Exercises

In order to assess the impact of readability and lexical richness of programming problem descriptions on students' cognitive load, we considered several metrics. These included the time taken to solve each problem and a self-assessment of cognitive load using the NASA-TLX questionnaire at the end of each task [5].

The readability of each problem description was measured using the Flesch Reading Ease index, and lexical richness was assessed using indices such as the Type-Token Ratio (TTR). Problem descriptions were categorized as low, medium, or high readability and lexical richness.

Low Readability and Lexical Richness: Descriptions with unnecessary technical vocabulary and long sentences. **Medium readability and lexical richness:** Clear descriptions with appropriate technical language. **High readability and lexical richness:** Detailed descriptions including clear examples and expected results.

4.1   Exercises

Below, we present an exercise adapted for three levels. The low-level exercise has a Flesch Reading Ease score 9.32, categorizing it as professional and extremely difficult to read. In contrast, the medium-level exercises scored 35.95, and the high-level exercise scored 37.23, both of them fall under the College level, classified as difficult to read.

Low-Level JavaScript Exercise: Description: Given a set of positive numbers in a sequential structure, the operator must perform an iteration to determine the maximum accumulated value, taking into account the present numbers. For each number, it should be compared with the previous ones to maintain the highest at each step and store it for later printing. **Requirements:** Use a loop to iterate over the elements of the set. Compare each value with the highest found so far. Display the maximum value. **Flesch Reading Ease index:** 9.32

Medium-Level JavaScript Exercise: Description: Write a JavaScript program that receives an array of positive integers and determines the largest number in the array. **Requirements:** Define an array of integers. Iterate through the array using a loop. Compare each number with the highest value found so far. Print the largest number at the end of the execution. **Flesch Reading Ease index:** 35.95.

High-Level JavaScript Exercise: Description: Write a JavaScript program that, given an array of integers, finds and prints the largest number in the array. The program must iterate over the array elements and keep track of the largest number found so far. **Requirements:** Declare an array with several integers. Iterate through the array using a for loop. Compare each number with the largest value found so far and update the largest if necessary. Print the largest number at the end. **Example Input:** let numbers = [3, 7, 2, 9, 5]; **Expected Output:** The largest number is: 9 **Flesch Reading Ease index:** 37.23.

4.2 NASA-TLX Questionnaire for Evaluating Cognitive Load in Programming Tasks

The NASA-TLX (Task Load Index) is a commonly used tool to assess cognitive load in various tasks. The questionnaire focuses on six key dimensions: mental demand, physical demand, temporal demand, performance, effort, and frustration level [5]. In Table 2 we show the adapted questions for evaluating cognitive load while performing a programming task.

Since that the task involves programming, we can assign different weights to the NASA-TLX dimensions, with physical demand being the least important. With Mental demand: 30%, Physical demand: 5% Temporal demand: 15%, Performance: 20%, Effort: 20% and Frustration: 10%. These weights reflect the importance of mental effort, performance, and effort during programming tasks, while reducing the emphasis on physical activity.

5 Exploratory Analysis

In this study, students were asked to solve two programming problems with descriptions of different readability and lexical richness, presented in random

Table 2. NASA-TLX Questionnaire for Programming Tasks with a Scale of Very Dissatisfied (1) to Very Satisfied (10)

Dimension/Question
Mental Demand: How much mental and perceptual activity was required for the programming task (thinking, deciding, calculating, remembering, searching for information)?
Physical Demand: How much physical activity was required to complete the task (typing code, using the keyboard, navigating through the code)?
Temporal Demand: How pressured did you feel by time while solving the task?
Performance: How satisfied are you with your level of performance upon completing the task?
Effort: How much mental and physical effort did you have to put forth to complete the programming task?
Frustration Level: How frustrated, insecure, stressed, or irritated did you feel during the task?

order. The first problem involved basic-medium JavaScript, and the second involved constructing a component with React.

The experiment was conducted using the Canvas platform, which allows tracking the start and end times of each programming exercise. Additionally, after completing the task, students answered the questionnaire to assess their perceived cognitive load in dimensions such as mental effort, mental demand, temporal demand, and frustration [8]. Subsequently, the professor graded the exercise and assigned a grade on a scale from 1 to 5, where 5 indicates the highest grade, meaning the student completed the exercise, and 1 indicates that the student only completed part of the exercise structure. As shown in Fig. 1, we present the process followed in the experiment, which begins with a random assignment of readability level for the student, continues with the completion of the JavaScript and React exercises, and ends with the professor's grade assignment.

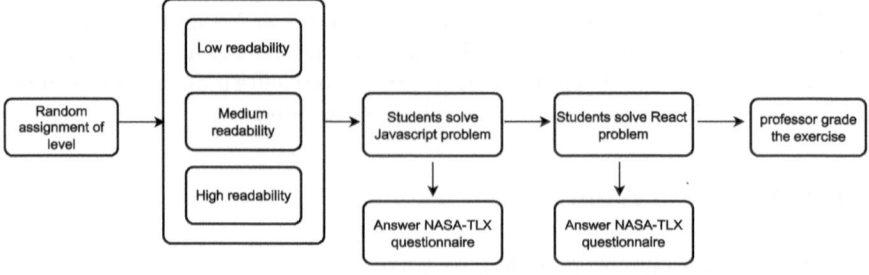

Fig. 1. Process of the Exploratory Analysis

Table 3 shows the time taken by each student, their grade, the level of readability and lexical richness of the problem description, and their cognitive load as measured by the NASA-TLX questionnaire.

Table 3. Results of the first programming exercise with cognitive load (NASA-TLX)

Student	Time (min)	Grade	Readability Level	NASA-TLX (CL)
Student1	15	5	Low	3.9
Student2	12	5	Low	3.8
Student3	34	2	Low	5.2
Student4	15	5	Medium	5.3
Student5	22	5	Medium	6.0
Student6	30	1	Medium	7.9
Student7	32	5	High	3.5
Student8	32	1	High	4.9
Student9	16	3	High	6.3

The NASA-TLX cognitive load scores provides information about how students experienced the task. For the level of Low Readability/Lexical Richness, we observed Student1 and Student2 performed well in terms of time and score, with low cognitive load (3.9 and 3.8). Student3, who took the longest time (34 min), reported a higher cognitive load (5.2), indicating confusion with the task. For the level of Medium Readability/Lexical Richness, both Student4 and Student5 scored full marks, but Student4 reported a slightly lower cognitive load (5.3 vs. 6.0). Student6 struggled significantly, with a high cognitive load (7.9) and poor performance, suggesting a lack of fundamental programming skills. And for High Readability/Lexical Richness, Student7 handled the more complex instructions well, reporting a low cognitive load (3.5). In contrast, Student8 struggled with a higher cognitive load (4.9) and a low grade, possibly due to insufficient programming experience. Student9 also exhibited a higher cognitive load (6.3) and encountered logical errors.

Cognitive load, as measured by NASA-TLX, appears influenced by both the readability and lexical richness of the task description, as well as the student's programming proficiency. Students with a stronger foundation (e.g., Student1, Student2, Student7) reported lower cognitive loads and performed better, even with complex instructions. In contrast, students with weaker skills (e.g., Student6, Student8) experienced higher cognitive loads and poorer outcomes. These findings suggest that while readability and lexical richness impact performance, their effect is mediated by programming ability. Simplifying descriptions for less proficient students may reduce cognitive load and improve performance, whereas increased complexity does not negatively impact more skilled students.

In the second programming task, students were asked to create a React component using JavaScript. Table 4 summarizes the time spent, grades, NASA-TLX cognitive load, and the readability and lexical richness of the task descriptions.

Table 4. Results of the second programming task with cognitive load (NASA-TLX) scores

Student	Time (min)	Grade	NASA-TLX (CL)	Readability Level
Student1	24	5	4.65	Low
Student2	27	5	6.45	Low
Student3	25	4	6.35	Low
Student4	83	5	6.35	Medium
Student5	45	1	8.2	Medium
Student6	43	5	7.75	Medium
Student7	31	4	4.55	High
Student8	31	2	5.5	High
Student9	59	4.5	8.15	High

Cognitive load in this task was influenced by both readability/lexical richness and the student's approach. Students with lower cognitive loads (e.g., Student1, Student7) generally performed better, while those with higher cognitive loads (e.g., Student5, Student9) struggled, often due to difficulties following instructions or taking excessive time.

Additionally, time spent on the task did not consistently correlate with performance or cognitive load, as seen with Student4, whose high time did not lead to higher cognitive load or lower performance.

6 Comparison of Performance Between the First and Second Programming Task

In this section, we compare the results of the students in the two different programming exercises detailed above, where both tasks were evaluated similarly.

In Table 5, we observe most students spent more time on the second task compared to the first one, with Student4, Student5, Student6, and Student9 showing a significant increase in the time required. The nature of the second task, which involved building a React component, likely contributed to this increase in time due to its complexity compared to the simpler JavaScript program in the first task. We also note the performance was relatively consistent for some students (e.g., Student1, Student2), who maintained high grades (5) across both tasks. However, other students like Student5, who received a grade of 5 in the first task, dropped to 1 in the second task, reflecting difficulties in following instructions

Table 5. Combined results of the first and second programming tasks with cognitive load (NASA-TLX)

Student	First Programming Task			Second Programming Task		
	Time (min)	Grade	NASA-TLX	Time (min)	Score	NASA-TLX
Student1	15	5	3.9	24	5	4.65
Student2	12	5	3.8	27	5	6.45
Student3	34	2	5.2	25	4	6.35
Student4	15	5	5.3	83	5	6.35
Student5	22	5	6.0	45	1	8.2
Student6	30	1	7.9	43	5	7.75
Student7	32	5	3.5	31	4	4.55
Student8	32	1	4.9	31	2	5.5
Student9	16	3	6.3	59	4.5	8.15

or the increased complexity of the React task. Student8, who also struggled in both tasks, improved slightly but got again a second low grade.

Students with higher cognitive load generally performed worse for the Cognitive Load in both tasks. For example, Student 5 experienced a significant cognitive load increase (from 6.0 to 8.2), which correlated with his lower grade in the second task. Similarly, Student 6 maintained a high cognitive load in both tasks, though his performance improved significantly in the second task, suggesting that he handled the complexity better despite the difficulty. Student1 and Student7 had the lowest cognitive load in the second task, corresponding to their ability to complete the task more efficiently.

The comparison between the two programming tasks highlights how increased task complexity, as seen in the React component exercise, influenced both the time spent and cognitive load. Students who exhibited strong programming skills in the first task (e.g., Student1, Student2) could maintain their performance in the second task, even as cognitive demands increased. On the other hand, students like Student5 and Student8 struggled significantly more with the increased complexity, reflected in their lower grades and higher cognitive load.

7   Conclusion

The experiment suggests that the readability of problem descriptions influences students' performance in programming tasks. Those with lower programming experience struggled with more complex language, resulting in higher cognitive loads and poorer performance. On the other hand, students with more experience demonstrated greater resilience to increased task complexity, reporting lower cognitive load and better performance.

These findings support the idea that instructional material should be adapted to the students' skill level. By carefully managing cognitive load, teachers can optimize learning and avoid overwhelming students. Future research could explore the effect of more dynamic adjustments to readability and complexity based on real-time cognitive load measurements. We also foresee an intelligent learning environment where a learner profile handling information about his/her level can be matched with the corresponding programming exercises. As the student advances in programming skills, the assigned exercises can be more challenging, with higher readability.

Acknowledgement. Authors were partially supported by SNII-Conahcyt.

References

1. Baddeley, A.D., Hitch, G.J.: Working memory. Psychol. Learn. Motiv. **8**, 47–89 (1974)
2. Berssanette, J.H., de Francisco, A.C.: Cognitive load theory in the context of teaching and learning computer programming: a systematic literature review. IEEE Trans. Educ. **65**(3), 440–449 (2022)
3. Eleyan, D., Othman, A., Eleyan, A.: Enhancing software comments readability using Flesch reading ease score. Information **11**(9), 430 (2020)
4. Gogna, Y., Tiwari, S., Singla, R.: Evaluating the performance of the cognitive workload model with subjective endorsement in addition to EEG. Med. Biol. Eng. Comput. (2024)
5. Hart, S.G., Staveland, L.E.: Development of NASA-TLX (task load index): results of empirical and theoretical research. In: Advances in Psychology, vol. 52, pp. 139–183. Elsevier (1988)
6. Kirschner, P.A., Sweller, J., Clark, R.E.: Why minimal guidance during instruction does not work: an analysis of the failure of constructivist, discovery, problem-based, experiential, and inquiry-based teaching. Educ. Psychol. **41**(2), 75–86 (2006)
7. Sweller, J.: Cognitive Load Theory: Recent Theoretical Advances, pp. 29–47. Cambridge University Press (2010)
8. Sweller, J.: Cognitive load theory. Psychol. Learn. Motiv. **55**, 37–76 (2011)
9. Van Merriënboer, J.J., Sweller, J.: Cognitive load theory in health professional education: design principles and strategies. Med. Educ. **44**(1), 85–93 (2010)

CIAPP 2024

Air Pollution, Socioeconomic Status, and Avoidable Hospitalizations in Mexico City: A Multifaceted Analysis

Carlos Minutti-Martinez[1]($\boxtimes$) , Miguel F. Mata-Rivera[2] ,
Magali Arellano-Vazquez[1] , Boris Escalante-Ramírez[3] ,
and Jimena Olveres[3]

[1] INFOTEC Centro de Investigación e Innovación en Tecnologías de la Información y
Comunicación, Aguascalientes, Mexico
{carlos.minutti,magali.arellano}@infotec.mx
[2] UPIITA, Instituto Politécnico Nacional, Mexico City, Mexico
mmatar@ipn.mx
[3] Center for Advanced Computing Studies, National University of Mexico,
Mexico City, Mexico
{boris,jolveres}@cecav.unam.mx

Abstract. Previous studies have independently examined the relationships between air pollutants, socioeconomic status (SES), and disease incidence, often neglecting to consider both risk factors simultaneously. However, evidence suggests a correlation between SES and air pollution levels, implying their potential confounding effects. Failing to account for this could lead to inaccurate conclusions. Moreover, SES analyses frequently focus solely on income or residential area, overlooking crucial factors like educational attainment. Air pollution studies tend to concentrate on specific pollutants such as PM2.5 or PM10, disregarding other potentially relevant, highly correlated pollutants, potentially misidentifying the most influential ones. Incorporating multiple risk factors introduces complexities like multicollinearity, which can distort effect estimates and statistical significance.

In this study, we analyzed 86,170 hospitalized patients from Mexico City between 2015 and 2019, utilizing composite social and economic indicators. We included various environmental pollutants to comprehensively examine their contributions and effects on both the number and severity of hospitalizations. The results showed that the economic factor of SES significantly influenced the incidence of diabetes complications, while the social factor impacted diseases related to prenatal care and hypertension. The PM10, PM2.5, and CO pollutant group had a statistically significant effect on the incidence of several conditions like influenza, asthma, and epilepsy. The NO2 and NOx group exhibited effects on the severity of diabetes complications and influenza. Notably, nonlinear effects and interactions between variables like age and weight were observed, highlighting the importance of accounting for these complexities.

© The Author(s), under exclusive license to Springer Nature Switzerland AG 2025
L. Martínez-Villaseñor et al. (Eds.): MICAI 2024 Workshops, LNAI 15465, pp. 73–86, 2025.
https://doi.org/10.1007/978-3-031-83882-8_8

Keywords: Avoidable hospitalizations · Ambulatory care sensitive conditions · Air pollution · Socioeconomic status · Environmental health · Epidemiology · Risk factors · Mexico City

1   Introduction

Avoidable hospitalizations (AH), also known as ambulatory care sensitive conditions (ACSCs), refer to hospital admissions potentially preventable through timely and effective outpatient care. These are often associated with chronic illnesses like diabetes, asthma, and congestive heart failure, as well as acute conditions such as pneumonia and complicated appendicitis. Effective primary care can help prevent or manage these conditions, reducing hospitalization needs [8,14,15].

Avoidable hospitalizations are a key indicator of primary care quality, frequently resulting from inadequate or delayed community-based care. Their occurrence highlights the need for better care coordination, preventive services, and disease management strategies across healthcare settings [8,14].

Lower socioeconomic status (SES) is associated with a higher risk of avoidable hospitalizations, which could have been prevented through timely outpatient care [2,17,20]. The combined effect of individual-level household income and neighborhood-level material deprivation on hospitalization risk is significant, with those living in low-income neighborhoods experiencing the highest risk. While the mechanisms are not fully understood, factors like limited healthcare access, health behaviors, and health outcomes are thought to play a role [20].

Lower-SES populations are often disproportionately exposed to higher air pollution levels, contributing to increased mortality risks from all causes, including respiratory causes [1,6]. Despite extensive air pollution epidemiology research recognizing this disparity, there is limited understanding of optimal SES confounding adjustment methods and potential biases arising from improper adjustment [6].

Due to the high correlation between social and economic factors, determining their relative contributions is challenging. Consequently, SES analyses often focus primarily on income or residential area, neglecting crucial factors like educational attainment.

Avoidable hospitalization rates vary considerably across countries, including Mexico. For instance, asthma admission rates vary 12-fold across OECD countries, with Mexico, Italy, and Colombia reporting the lowest rates, while Latvia, Turkey, and Poland report rates more than twice the OECD average [11].

Understanding avoidable hospitalizations in specific populations requires consideration of contextual factors such as socioeconomic status (SES), healthcare access, and health behaviors. By studying these populations, policymakers can identify necessary changes to address avoidable hospitalizations, ultimately improving healthcare outcomes and reducing costs.

Analyzing trends in avoidable hospitalizations by clinical condition helps inform healthcare policy and resource allocation by identifying increasing or decreasing rates over time. This examination can reveal patterns and correlations between specific conditions and hospitalization rates, guiding targeted interventions. Moreover, it can help identify high-risk groups based on age, gender, or socioeconomic status.

In this study, we investigate the principal risk factors for avoidable hospitalizations in Mexico City, examining their relationship with SES and air pollution, high-risk groups, and temporal changes. Our analysis of 86,170 patient records from 2015 to 2019 employed negative binomial regression, logistic regression, and Gradient Boosting Machine (GBM) models to account for non-linearity and interactions between variables. We included SES and air pollution as risk factors, along with relevant covariates such as locality, age, gender, weight, and admission date.

To examine SES effects, we generated a composite indicator using factor analysis (FA), distinguishing economic and social factor contributions. For air pollution, we constructed indexes through Principal Component Analysis (PCA) to account for spatial pollutant concentration correlations across localities. We applied an iterative algorithm tailored to the problem to obtain relevant factors, eliminating non-significant variables and penalizing the simultaneous inclusion of highly correlated variables to reduce multicollinearity and interpretation problems.

Our results show that different aspects of the composite SES indicator influence the incidence of various avoidable hospitalization categories, while environmental air pollution affects both the incidence and severity of hospitalizations. Notably, interactions and nonlinear effects between variables were found, which can inform prevention efforts and public policy to reduce avoidable hospitalizations.

This paper is structured as follows: an *Introduction* providing context on avoidable hospitalizations, socioeconomic status, and air pollution; a detailed *Methodology* outlining data sources, composite indicator construction, and negative binomial regression, logistic regression, and Gradient Boosting Machine models employed; a *Results* section presenting findings on statistically significant risk factors for each condition category, including nonlinear effects and interactions; and a concluding *Summary and Conclusions*, synthesizing key results and their implications for targeted interventions and policy refinements to reduce avoidable hospitalizations in Mexico City. This integrated approach elucidates the interplay between socioeconomic status, air pollution, and individual characteristics, informing evidence-based strategies to improve healthcare outcomes.

2   Methodology

To determine the relevance of air pollution (AP) and socioeconomic status (SES) on the leading causes of hospitalization, the patient's locality of residence is matched with other datasets to estimate the corresponding AP and

SES indexes. Multiple confounding factors are included, such as sex, age, weight, access to social security, municipality of residency, and admission date (months 1–60), among others.

Severity was determined by the number of hospitalization days and whether death occurred. Each variable's contribution was estimated using an iterative algorithm tailored to the problem, which eliminates and includes variables based on the Akaike Information Criterion (AIC). Additionally, to explore non-linearity and variable interactions, the relative feature importance from the Gradient Boosting Machine (GBM) model was utilized. All variables were scaled to 0–1 for comparison.

2.1   Data Sources

Three main data sources were used:

Hospitalizations. Anonymized data from public hospitals were provided by the Mexico City Ministry of Health (SEDESA) under CONACYT project 7051. The data included patient information such as age, weight, gender, origin, hospitalization indicator, health services entitlement, admission and discharge dates, hospitalization days, conditions, locality of residence, and International Classification of Diseases (ICD) codes for initial diagnosis, main condition, and cause of death.

Air Pollutant Concentrations. AP measures were obtained from Mexico City's Automatic Air Quality Monitoring Network [13]. A 15-year average (2005–2020) was calculated for each station's concentrations of PM_{10}, $PM_{2.5}$, CO, NO_X, NO_2, SO_2, NO, and O_3. Patient locality mean concentrations were obtained by kriging and QGIS, using the locality centroid.

Census of Population and Housing. The official 2020 Mexico Census data set contained housing and population variables at the locality level, used to construct SES indicators.

2.2   SES and AP Factors

Census data are widely used for neighborhood-level composite SES indicators, utilizing Principal Component Analysis (PCA) or Factor Analysis (FA) to weigh each variable's contribution [9, 21].

This study derived SES indicators using FA to separate economic and social factors (F_ECONOM and F_SOCIAL), with higher values representing less favorable circumstances. These factors were validated by regressing them against the Social Gap Index (SGI) [3], Social Development Index (SDI) [4], and Human Development Index (HDI) [18], resulting in coefficients of determination $R^2 > 0.9$. The Economic factor (F_ECONOM) was more correlated with housing variables in the Census, like the number of dwellings with latrine, with only one room, or dwellings with earthen floors. Meanwhile, the Social factor

(F_SOCIAL) was more correlated with variables like the average number of live-born daughters and sons, average level of schooling, and affiliation to different health services.

Regarding AP, Mexico City's complex terrain influenced meteorology and pollutant behavior in the atmosphere, resulting in spatially correlated AP. Failing to account for this could lead to incorrect pollutant effect estimates. Therefore, factors were constructed by grouping highly correlated pollutants by location. PCA was used to determine which pollutants to group due to their correlated concentrations by location, resulting in the three groups: PM_CO (PM_{10}, $PM_{2.5}$, & CO), NO2_NOx (NO_X & NO_2), and SO2_NO_O3 (SO_2, NO, & O_3).

2.3 Models

Hospitalization severity for each patient was measured as a composite indicator expressing death occurrence and hospitalization days, where the severity Y of patient i had $Y_i > 0.5$ when death occurred, and the closer to 1, the faster the death, indicating greater severity. If $Y_i < 0.5$, death did not occur, and the closer to 0, the fewer hospitalization days before discharge, indicating less severity. This formulation also can be interpreted as a classification problem with high (death) and low (non-death) severity classes, weighted for extreme cases. A more detailed explanation of how the SES and AP indicators were constructed, as well as how the severity was estimated, can be found in [10].

A compound catalog of Avoidable Hospitalization for Ambulatory Care Sensitive Conditions ICD 10 codes was extracted from [12]. Table 1 presents the 14 categories that had enough data to study severity and the number of cases by locality and admission date.

Table 1. Avoidable hospitalizations categories and number of cases analyzed

Code	Category	Records
DC	Diabetes complications	23868
I&P	Influenza and pneumonia	16075
EN&T INFEC	Ear, nose and throat infections	7146
GASTRO	Dehydration and gastroenteritis	5252
PYELO	Pyelonephritis	5218
ULCER	Perforated or bleeding ulcer	5172
CELL	Cellulitis	4428
ASTH	Asthma	4278
DPCPD	Diseases related with the prenatal health care of pregnancy and delivery	4004
EPILEP	Convulsions and epilepsy	3279
HYPERT	Hypertension	2319
HEART	Congestive heart failure	2110
COPD	Chronic obstructive pulmonary disease	1600
ANG	Angina	1421

Two types of models were of interest in this study: (1) models to estimate the monthly number of hospitalizations of patients with specific conditions by locality, and (2) models to estimate the severity of hospitalization.

In the case of the number of hospitalizations, at the locality level, the total population of the locality (POBTOT) was expected to be the main predictor. In addition, other predictors were included, such as the proportion of the population aged 0 to 2 years (P_0A2), the proportion of the population aged 18 to 24 years (P_18A24), the proportion of the population aged 60 years and over (P_60YMAS), population density (POB_AREA), Male-female ratio (REL_H_M), and the SES and AP factors presented above. At the patient level, the patient's municipality of residence (E_MUN_XXXXX), date of admission (ADM_DATE) expressed from month 1–60, and month of admission (MONTH) were considered.

For studies modeling the number of hospitalizations, Wallar *et al.* [19] performed a meta-analysis in which they concluded that the most appropriate model is a negative binomial regression, which is the model used in this study.

Logistic regression is widely regarded as an appropriate model for predicting mortality during hospitalization due to several key attributes that enhance its effectiveness in clinical settings (see [16]). At the patient level, the variables considered relevant for modeling severity were age (AGE), weight (WEIGHT), gender (SEX_M, 1 if male, 0 otherwise), and origin (PROCED, 1 if external, 2 for emergency, 3 for referred, 4 for other, 9 for unspecified). At the locality of residence level, the patient's municipality of residence (E_MUN_XXXXX) and date of admission (ADM_DATE) expressed as months 1–60 were considered.

In both models, the municipality of residence was relevant since each municipality may have different hospital infrastructure, health policies, or factors not considered that could present a spurious correlation with SES or the presence of AP.

Additionally, a variable selection algorithm was employed. It started by considering the $n = 10$ variables most correlated with the variable to be modeled. Then, in an iterative process, statistically non-significant variables were eliminated one by one, removing the variable with the highest p-value and adjusting the model again. If all variables were significant, the next most correlated variable was included. Ultimately, the model with the lowest $AIC + \lambda \cdot r$ was selected, where λ is a penalty value for the maximum absolute correlation among explanatory variables (r), aiming to reduce multicollinearity.

Conventional modeling methods often struggle to capture high-dimensional relationships effectively. In contrast, some advanced machine learning techniques like the Gradient Boosting Machine (GBM) model have been successfully applied in predictive analytics for medical applications, outperforming other ML and statistical models (*e.g.*, Kong *et al.* [7]). The GBM was also utilized to automatically account for nonlinear confounding effects and interactions, estimating the effect of each variable and exploring relationships that could be challenging to detect with more classic models.

For the analysis, 15% of records were set aside for validation for each category, ensuring that predictions were not random and that the variable importance measurements had predictive value.

3   Results

The results of the relevant factors obtained for each model are presented below. Standardized coefficients are reported so that it is possible to compare the magnitudes of the effects within each model. Statistical significance was considered for p-value ≤ 0.05. When a predictor variable does not appear in the list of relevant factors, it can be assumed that its effect was not statistically significant or that it was highly correlated with another variable whose effect was greater.

The effects due to the different municipalities are presented for contrast purposes. Red colors in the graphs represent effects that increase the number of hospitalizations or their severity, the blue color is the opposite situation.

In the case of the GBM model, the normalized Gini importance of each variable is estimated and the 10 variables with more importance are presented.

In Fig. 1 is presented the statistically significant risk factors for Diabetes complications on the number and severity of hospitalizations and its magnitude. It is shown that the patient's weight is one of the most significant factors increasing the severity and also the NO_2 and NO_x pollutants have an important contribution. Regarding the number of hospitalizations, as expected, the total population of the locality is the variable with the largest impact, but with a similarly large magnitude it can be found that the Economic Factor is important, having a larger number of hospitalizations those localities with less favorable economic status. Some localities of residence shown a specific effect, not explained by the other variables.

In Fig. 2 is presented a result using the GBM, showing the interaction between different pairs of variables and how it affects the number of hospitalizations and the severity. It can be seen the importance to analyze interactions and non linear effects, for example, admission date did not show to be statistically significant in the regression analysis, but in the GBM analysis it is shown a nonlinear effect where the number of cases is increasing up to month 30, and decreasing after that. This behavior can lead to a non significant effect on a linear model, even though it could have a significant contribution. Similarly, when the relationship between Social and Economic factors is studied, it can be seen that, as shown by the statistical model, the larger contribution is mainly by the economic factor, but, when the social factor is largely unfavorable, it also have an impact when it interacts with a large unfavorable economic factor.

Also from analyzing Fig. 2 and the effects on the severity of hospitalizations, it is shown a linear effect due to the patient's age, but also a nonlinear interaction between the patient's age and weight, where the larger severity is found for patients with high age, and with high or low weight. As determined with the logistic regression, the NO_2 and NO_x pollutants show an increase in severity, but it also can be seen an nonlinear interaction of larger severity when also PM and CO are present in the environment.

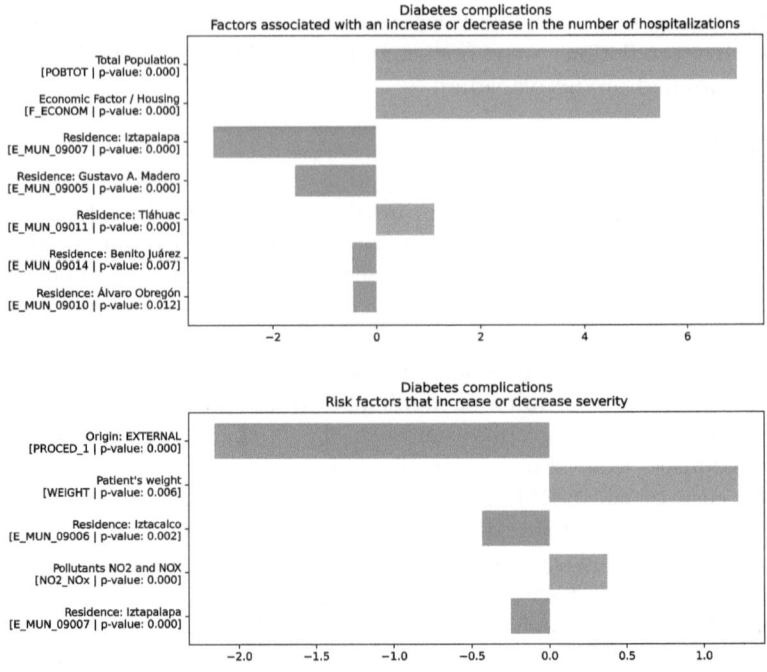

Fig. 1. Relevant factors on the number and severity of hospitalizations for Diabetes complications.

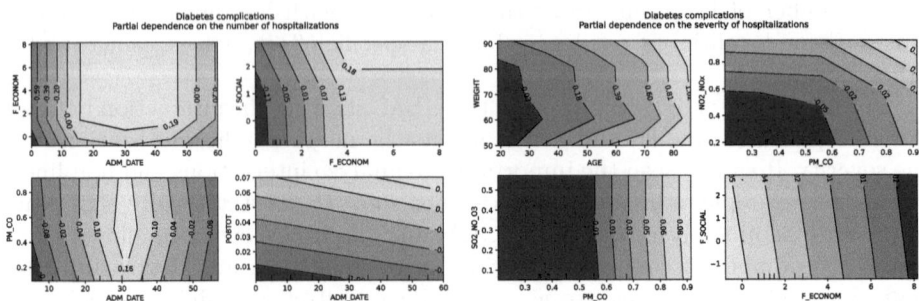

Fig. 2. Partial dependence on the number and severity of hospitalizations for Diabetes complications.

From these two sets of figures it can be seen how both models can be used to detect the most significant risk contributors to the severity and number of hospitalizations, taking into account also nonlinear effects and interactions between variables, which could be unknown.

Tables 2 and 4 present the variables of interest with have statistically significance to model the number and severity of hospitalizations for each of the avoidable hospitalizations categories studied, and the magnitude of the effect, after

Table 2. Variables with statistically significant effects on the number of avoidable hospitalizations, by category (Logistic Regression)

	POBTOT	ADM_DATE	F_ECONOM	F_SOCIAL	PM_CO	NO2_NOx	SO2_NO_O3
DC	6.979		5.493				
I&P	5.961				1.256		2.584
EN&T INFEC	4.474	−0.426			1.790		
GASTRO	1.616				1.451		
ULCER	3.768						
PYELO	2.404						
CELL	3.581						
ASTH	1.433				1.666		
DPCPD	2.647			2.252			
EPILEP	1.870				0.602		
HYPERT	2.234			0.520			
HEART	2.758						
ANG	2.499						
COPD	1.645						

Table 3. Normalized Gini importance of variables in predicting the number of avoidable hospitalizations, by category (Gradient Boosting Machine)

	POBTOT	ADM_DATE	F_ECONOM	F_SOCIAL	PM_CO	NO2_NOx	SO2_NO_O3
DC	**0.489**	0.126	**0.075**	0.005	0.013	0.010	0.087
I&P	**0.533**	0.046	0.007	0.088	**0.027**	0.003	**0.002**
EN&T INFEC	**0.530**	**0.080**	0.007	0.007	**0.170**	0.023	0.001
GASTRO	**0.670**	0.053	0.003	0.013	**0.105**	0.009	0.010
ULCER	**0.580**	0.099	0.026	0.018	0.031	0.004	0.003
PYELO	**0.765**	0.063	0.005	0.024	0.005	0.006	0.002
CELL	**0.745**	0.078	0.006	0.003	0.004	0.003	0.009
ASTH	**0.496**	0.084	0.003	0.004	**0.163**	0.083	0.002
DPCPD	**0.654**	0.066	0.021	**0.013**	0.008	0.008	0.039
EPILEP	**0.753**	0.058	0.003	0.004	**0.015**	0.005	0.018
HYPERT	**0.755**	0.076	0.009	**0.013**	0.004	0.007	0.034
HEART	**0.728**	0.087	0.010	0.009	0.005	0.022	0.009
ANG	**0.570**	0.162	0.005	0.006	0.010	0.008	0.014
COPD	**0.550**	0.132	0.013	0.018	0.031	0.008	0.023

*Bold text indicates statistical significance in the regression analysis.

applying the iterative variable selection algorithm. In a similar way Tables 3 and 5 present the normalized Gini importance for each variable estimated with the GBM model for number and severity of hospitalizations, marking in bold the variables with statistical significance in their regression counterpart. A comparative between both results show relevant insights, for example, although the Gini importance does not reveal the direction of the effect, when the importance is large, but was not statistically significant in the regression analysis it suggests a possible non-linear effect.

The results of the logistic model show that the total population (POBTOT) is the variable with the largest effect and its presence is consistent across each category, aligning with expected results. Regarding the importance of variables in the GBM model, POBTOT is also the most relevant variable. However, for variables like the admission date (ADM_DATE), only EN&T INFEC (Ear, nose and throat infections) shows a statistically significant effect, with a reduction in the number of cases over time. But when the results of using GBM are analyzed, it shows a large effect for categories like DC, ANG, and COPD. A possible explanation is the behavior already described in Fig. 2, where there is a nonlinear effect of increase and decrease over time that could potentially eliminate the overall effect in a linear model.

When the Economic factor (F_ECONOM) is analyzed, it is observed that statistically it only impacts the number of hospitalizations for Diabetes Complications (DC), and the GBM also found that DC is the category with the largest effect for this risk factor. For the Social factor (F_SOCIAL), the Diseases related with the prenatal health care of pregnancy and delivery (DPCPD) shows a large effect, where a less favorable value increases the number of hospitalizations in those localities. Hypertension (HYPERT) also shows an increase in hospitalizations for less favorable values of F_SOCIAL.

Environmental Air pollution also shows statistically significant effects for some categories. The group of PM_{10}, $PM_{2.5}$ & CO (PM_CO) increases the number of hospitalizations for Influenza and pneumonia (I&P), Dehydration and gastroenteritis (GASTRO), EN&T INFEC, Asthma (ASTH), and Convulsions & epilepsy (EPILEP). The group of SO_2, NO & O_3 (SO2_NO_O3) has a significant effect for I&P.

When the importance of variables produced by the GBM is analyzed, it shows similar results, where the PM_CO group of pollutants has the largest effects among the other groups, and ASTH and EN&T INFEC are the categories with the highest importance.

Regarding the severity of the hospitalizations (Tables 4 and 5), it demonstrates that Age and Weight have the largest effects on increasing severity. Age has statistically significant effects in nearly all categories, with Influenza and pneumonia (I&P) having the largest effect. Weight exhibits the largest effects on Diabetes Complications (DC) and Hypertension (HYPERT). Although Weight is not statistically significant in many categories, the GBM shows importance in numerous categories, indicating a possible nonlinear effect and even an interaction with Age. For instance, for some categories like Asthma (ASTH), there is no statistically significant effect, but it has a high effect in the GBM. When partial dependence was analyzed, it revealed that high and low weight increase severity, implying a nonlinear effect. Regarding sex, Pyelonephritis (PYELO) shows an increased severity for males, while Congestive heart failure (HEART) exhibits increased severity for females.

ASTH shows a decrease in severity for unfavorable Economic status and for the PM_CO group of pollutants, which could be biased due to the inability of sicker people to live in localities with high levels of pollution.

Table 4. Variables with statistically significant effects on the severity of avoidable hospitalizations, by category (Logistic Regression)

	AGE	WEIGHT	SEX_M	F_ECONOM	F_SOCIAL	PM_CO	NO2_NOx	SO2_NO_O3
DC		1.225					0.374	
I&P	4.555	0.971					0.334	
EN&T INFEC	4.033							
GASTRO	0.195							
ULCER	2.211							
PYELO	4.069		0.447					
CELL	3.792							
ASTH				−0.037		−0.016		
DPCPD				0.031				
EPILEP	0.200							
HYPERT	3.405	2.339						
HEART	1.615			−0.238				
ANG	1.977							
COPD	3.314							

Table 5. Normalized Gini importance of variables in predicting the severity of avoidable hospitalizations, by category (Gradient Boosting Machine)

	AGE	WEIGHT	SEX_M	F_ECONOM	F_SOCIAL	PM_CO	NO2_NOx	SO2_NO_O3
DC	0.181	**0.096**	0.002	0.004	0.006	0.017	**0.023**	0.002
I&P	**0.927**	**0.020**	0.001	0.002	0.005	0.008	**0.002**	0.003
EN&T INFEC	**0.742**	0.101	0.004	0.010	0.009	0.008	0.013	0.006
GASTRO	**0.431**	0.110	0.012	0.060	0.042	0.013	0.028	0.012
ULCER	**0.042**	0.027	0.001	0.013	0.013	0.005	0.014	0.002
PYELO	**0.760**	0.039	**0.032**	0.008	0.007	0.023	0.006	0.011
CELL	**0.437**	0.087	0.024	0.052	0.034	0.019	0.023	0.026
ASTH	0.166	0.211	0.004	**0.016**	0.078	**0.043**	0.023	0.049
DPCPD	0.333	0.365	0.000	**0.026**	0.025	0.030	0.017	0.013
EPILEP	**0.388**	0.187	0.013	0.022	0.041	0.031	0.023	0.023
HYPERT	**0.359**	**0.255**	0.022	0.032	0.008	0.023	0.059	0.006
HEART	**0.274**	0.138	**0.014**	0.021	0.026	0.031	0.007	0.021
ANG	**0.277**	0.067	0.029	0.043	0.023	0.032	0.020	0.028
COPD	**0.469**	0.174	0.020	0.025	0.019	0.044	0.026	0.028

*Bold text indicates statistical significance in the regression analysis.

The group of air pollutants NO_2 and NO_x (NO2_NOx) presents an increase in severity for DC and I&P.

4   Discussion

This study's multifaceted approach, combining composite indicators for SES and air pollution with advanced statistical techniques, has unveiled complex interactions between risk factors for avoidable hospitalizations in Mexico City. By distinguishing between economic and social factors of SES and their differential impacts on various health outcomes, our findings emphasize the need for nuanced, targeted interventions in public health policy.

Our results highlight the importance of considering nonlinear effects and interactions between variables, particularly age and weight, in health outcome analyses. These complex relationships may explain why some risk factors, such as admission date, showed significant effects in the GBM model but not in regression. This underscores the necessity for sophisticated analytical approaches in epidemiological research to capture the full complexity of health determinants.

The observed differences in the impact of air pollution and SES on both the incidence and severity of avoidable hospitalizations provide a more nuanced understanding compared to previous studies. While numerous studies have independently examined the effects of air pollution or SES on health outcomes [5,19], our approach of simultaneously considering both factors reveals a more complex picture. For instance, we found that SES factors primarily influenced the number of hospitalizations, particularly for conditions like diabetes complications and prenatal care-related issues, while air pollution often affected both incidence and severity, as seen on diabetes complications and influenza. This distinction may be attributed to the chronic nature of socioeconomic disparities affecting long-term health behaviors and access to preventive care, whereas air pollution can have both acute and chronic effects on health.

5   Summary and Conclusions

Avoidable hospitalizations (AH) for ambulatory care-sensitive conditions (ACSC) have been considered an indicator of primary care performance in many countries. This study aimed to investigate how socioeconomic status (SES), environmental air pollution, and other factors contribute to increasing the incidence and severity of AH in the Mexican population. The findings provide insights into preventing these hospitalizations and guiding public policy focus areas. By utilizing a composite SES indicator, we were able to study the contribution of economic and social aspects, as well as the impact of different pollutant groups. From these analyses, we found the following key results:

- Social and economic factors have a larger impact on the number of AH than on the severity. The economic factor specifically affects the number of cases for Diabetes complications, while the social factor influences Diseases related to prenatal health care of pregnancy and delivery, as well as Hypertension.
- The group of air pollutants PM_{10}, $PM_{2.5}$ & CO has a statistically significant effect on the incidence of many AH categories, including Influenza and pneumonia, Ear, nose and throat infections, Dehydration and gastroenteritis, Asthma, and Convulsions and epilepsy.
- The group of pollutants SO_2, NO & O_3 shows an effect on the incidence of Influenza and pneumonia.
- NO_2 and NO_x exhibit an effect on the severity of Diabetes complications as well as Influenza and pneumonia.

- Multiple nonlinear effects and interactions between variables are present in AH, such as those between Age and Weight. These effects need to be studied in greater detail.

Although this work includes multiple confounding factors to assess the impact of SES and air pollution more accurately, further analyses are needed to better understand the interactions and complex behavior between different variables. Additionally, studying a larger population is necessary to better characterize how the examined factors contribute to AH in the Mexican population.

Data Availibility Statement. The code to generate the SAS indicators, contaminant factors and predictive models are openly available at https://github.com/cminuttim.

References

1. Blanco-Becerra, L.C., Miranda-Soberanis, V., Barraza-Villarreal, A., Junger, W., Hurtado-Díaz, M., Romieu, I.: Effect of socioeconomic status on the association between air pollution and mortality in Bogota, Colombia. Salud Publica Mex. **56**(4), 371–378 (2014)
2. Blustein, J., Hanson, K., Shea, S.: Preventable hospitalizations and socioeconomic status. Health Aff. (Millwood) **17**(2), 177–189 (1998)
3. CONEVAL: Índice de Rezago Social (IRS), 2020 (2021). https://www.coneval.org.mx/Medicion/IRS/Paginas/Indice_Rezago_Social_2020.aspx. Accessed 17 July 2022
4. EvaluaCDMX: Índice de Desarrollo Social de la Ciudad de México, 2020 (2021). https://evalua.cdmx.gob.mx. Accessed 17 July 2022
5. Hajat, A., Hsia, C., O'Neill, M.S.: Socioeconomic disparities and air pollution exposure: a global review. Curr. Environ. Health Rep. **2**(4), 440–450 (2015). https://doi.org/10.1007/s40572-015-0069-5
6. Hajat, A., MacLehose, R.F., Rosofsky, A., Walker, K.D., Clougherty, J.E.: Confounding by socioeconomic status in epidemiological studies of air pollution and health: challenges and opportunities. Environ. Health Perspect. **129**(6), 65001 (2021)
7. Kong, G., Lin, K., Hu, Y.: Using machine learning methods to predict in-hospital mortality of sepsis patients in the ICU. BMC Med. Inform. Decis. Mak. **20**(1), 251 (2020)
8. Lyhne, C.N., Bjerrum, M., Riis, A.H., Jørgensen, M.J.: Interventions to prevent potentially avoidable hospitalizations: a mixed methods systematic review. Front. Public Health **10**, 898359 (2022)
9. Messer, L.C., et al.: The development of a standardized neighborhood deprivation index. J. Urban Health **83**(6), 1041–1062 (2006). https://doi.org/10.1007/s11524-006-9094-x
10. Minutti-Martinez, C., Galindo, A., Valdez-Garduno, L.F., Mata-Rivera, M.F.: Exploring nonlinear effects of air pollution on hospital admissions by disease using gradient boosting machines. In: 2022 19th International Conference on Electrical Engineering, Computing Science and Automatic Control (CCE). IEEE (2022)
11. OECD: Avoidable hospital admissions (2019). https://doi.org/10.1787/59e3aa01-en

12. Poblano Verástegui, O., Torres-Arreola, L.D.P., Flores-Hernández, S., Nevarez Sida, A., Saturno Hernández, P.J.: Avoidable hospitalization trends from ambulatory Care-Sensitive conditions in the public health system in méxico. Front. Public Health **9**, 765318 (2021)
13. RAMA: Automatic Air Quality Monitoring Network (2022). http://www.aire.cdmx.gob.mx/default.php?opc=%27aKBh%27. Accessed 17 July 2022
14. Rosano, A., et al.: The relationship between avoidable hospitalization and accessibility to primary care: a systematic review. Eur. J. Pub. Health **23**(3), 356–360 (2012). https://doi.org/10.1093/eurpub/cks053
15. Sanderson, C., Dixon, J.: Conditions for which onset or hospital admission is potentially preventable by timely and effective ambulatory care. J. Health Serv. Res. Policy **5**(4), 222–230 (2000)
16. Shipe, M.E., Deppen, S.A., Farjah, F., Grogan, E.L.: Developing prediction models for clinical use using logistic regression: an overview. J. Thorac. Dis. **11**(Suppl 4), S574–S584 (2019)
17. Spycher, J., et al.: Potentially avoidable hospitalizations and socioeconomic status in Switzerland: a small area-level analysis. Health Policy **139**(104948), 104948 (2024)
18. UNDP: Informe de Desarrollo Humano Municipal 2010–2015 (2019). https://www.undp.org/es/mexico/publications/idh-municipal-2010-2015. Accessed 17 July 2022
19. Wallar, L.E., De Prophetis, E., Rosella, L.C.: Socioeconomic inequalities in hospitalizations for chronic ambulatory care sensitive conditions: a systematic review of peer-reviewed literature, 1990–2018. Int. J. Equity Health **19**(1), 60 (2020). https://doi.org/10.1186/s12939-020-01160-0
20. Wallar, L.E., Rosella, L.C.: Individual and neighbourhood socioeconomic status increase risk of avoidable hospitalizations among Canadian adults: a retrospective cohort study of linked population health data. Int. J. Population Data Sci. **5**(1) (2020). https://doi.org/10.23889/ijpds.v5i1.1351. https://ijpds.org/article/view/1351
21. Yu, M., Tatalovich, Z., Gibson, J.T., Cronin, K.A.: Using a composite index of socioeconomic status to investigate health disparities while protecting the confidentiality of cancer registry data. Cancer Causes Control **25**(1), 81–92 (2014). https://doi.org/10.1007/s10552-013-0310-1

Automatic Detection of Abnormal Pedestrian Flows, Using Classification and Tracking with Pre-trained YOLOv8

Adrián Núñez-Vieyra$^{(\boxtimes)}$ (ID), Rogelio Ferreira-Escutia (ID), Juan C. Olivares-Rojas (ID), Arturo Méndez-Patiño (ID), José A. Gutiérrez-Gnecchi (ID), and Enrique Reyes-Archundia (ID)

Instituto Tecnológico de Morelia, TecNM, Morelia, Mexico
{adrian.nv,rogelio.fe,juan.or,arturo.mp,jose.gg3,
enrique.ra}@morelia.tecnm.mx

Abstract. The use of artificial intelligence is a practice that is increasingly integrated into video surveillance systems, whether closed-circuit television systems or systems with intelligent IP cameras. In this way, video surveillance systems usually offer, at low cost, different functionalities such as object and person detection, facial recognition, issuing intrusion alerts, etc. However, with this dynamic has also increased the need to store, transmit and manipulate large amounts of data associated with video and the additional information that is generated daily. In this article we present a methodology and a video surveillance application focused on the automatic detection of abnormalities in pedestrian flow, using smart IP cameras, pre-trained YOLOv8 and statistical event recording with MongoDB. The accuracy tests in counting and tracking pedestrian flow that we have carried out give us results of Precision = 90%, Recall = 84%, Specificity = 62.5% and F1 = 87%, in addition to generating alerts when it detects abnormal pedestrian flows.

Keywords: Abnormal Pedestrian Flow · Computer Vision · Pedestrian Counting · Pedestrian Detection · Pedestrian Tracking · Video Surveillance · YOLO

1 Introduction

A video surveillance system, also called closed circuit television (CCTV), is defined as a technological tool that, through strategically located and interconnected video cameras, supports police operations and deployment, emergency response, crime prevention and the administration of justice [1]. Over the past two decades, video surveillance systems have evolved from generating useless databases to becoming intelligent systems capable of detecting, classifying and tracking objects and people across multiple cameras, solving authentication issues, and more. To achieve this automatic understanding, CCTVs carry out several processes that together are known as video analytics.

However, there are different challenges for video surveillance systems to achieve automatic image understanding: low-quality image captures, captures with inadequate

L. Martínez-Villaseñor et al. (Eds.): MICAI 2024 Workshops, LNAI 15465, pp. 87–98, 2025.
https://doi.org/10.1007/978-3-031-83882-8_9

angles, low processing capacity of the cameras, identification of recording sources (when there are several cameras), among others. In [2] it is mentioned that monitoring with real-time video surveillance systems still depends largely on the intervention of people, which implies human error and makes these systems less efficient, but there are still many scenarios where decision making is still faster and more accurate if there are human watching, compared to any type of automated video analysis [3], and this heterogeneity is one of the main challenges for automatic monitoring of video surveillance systems [4], for example, images may come from indoor or outdoor environments, have few or many people, include vehicles, pedestrians, animals, etc., or even the cameras may not be fixed, for example, if they are mounted on a service robot. Another problem is storage, for which there are already efforts to generate video synopsis, which is a technology that allows long videos to be shortened by projecting in a single frame activity of objects that originally occur at different times [3, 5–7]. In this paper we implement an automatic video surveillance application, capable of generating alerts when detecting abnormal pedestrian flows. An abnormal pedestrian flow occurs when the video surveillance system detects, in a specific time interval, that the number of people who have circulated in front of its cameras exceeds the expected values based on the historical statistics of the place or even if the pedestrian flow is abnormally low. To do this, we use smart IP cameras for video capture, a pre-trained version of YOLOv8 from Ultralytics to detect, classify and track people in real time, as well as an event logging process, using MongoDB, to generate a historical database that allows us to establish normal pedestrian flow patterns 24/7 in different scenarios. In general terms, the main contributions of this paper are as follows:

a) An application to automatically detect abnormal pedestrian flows in real time, based on machine learning and statistical data automatically generated for a specific environment.
b) A strategy is proposed to measure the accuracy of pedestrian counting, using a confusion matrix based on the uniqueness or duplication in the assignment of an identifier by a pre-trained tracking system.

2 Background

According to data from INEGI [8], Table 1 shows the crime rate per hundred thousand inhabitants reported in the last ten years in Mexico, highlighting street robbery, vehicle theft and home robbery. It is interesting to observe two things: the first is that these crimes begin and end in public roads and the second is that these crimes have decreased in recent years, which coincides with the growth and cheapening of technologies associated with video surveillance in the same period, this seems to indicate that the technology applied to video surveillance in public roads is having positive results.

Nowadays, in addition to their lower cost, video surveillance systems are capable of encoding, compressing and transmitting video to cloud servers, where more complex video analysis tasks are performed than those that each camera could perform individually [9]. But with these dynamic new problems arise, since it is now necessary to send more video over the Internet for processing, which implies acquiring connections and resources with higher performance. In [10] another problem associated with video surveillance is described, since now communications between capture and processing devices are often highly vulnerable to different attacks, taking into account the large volumes of video that must be transmitted to the cloud for later analysis. To solve this abundant flow of data, solutions such as video summarization are proposed [3, 5], or other alternatives such as decentralization of processing through more robust devices at the edge or periphery, which is known as fog computing [10].

Table 1. Crime incidence rate [8], highlighting crimes committed on public roads, as well as their decrease in recent years, which coincides with the growth of video surveillance systems.

	2013	2014	2015	2016	2017	2018	2019	2020	2021
Street robery	12,294	11,903	9,995	9,599	11,081	10,775	9,091	6,899	6,582
Extortion	9,790	9,850	8,600	8,945	7,719	6,542	5,134	5,160	5,375
Vehicle theft	4,218	4,213	3,457	3,611	3,755	3,645	3,132	2,718	2,803
Fraud	3,981	4,255	3,906	4,656	5,341	5,397	5,089	5,904	5,907
Verbal threats	3,808	4,109	2,835	2,872	3,323	3,253	3,090	2,958	2,823
Home robbery	2,689	2,534	2,496	2,437	2,745	2,598	2,063	1,880	1,849

3   Theoretical Framework

3.1   Video Analysis Process

In [9], five stages are described to carry out the video analysis process in a video surveillance system: detection, classification, tracking, interpretation and human identification. The detection stage aims at identifying regions of interest (ROIs) in images. ROIs are usually labeled as "foreground" and the rest of the image as "background". There are different algorithms to detect ROIs: background difference methods, optical flow methods, and frame difference methods. For example, in [3] a simplified algorithm called background-cut is used to separate the foreground from the background. On the other hand, in [11], they use a method that detects objects based on color or shape. Another technique to separate the background from the foreground is based on adaptive gaussian learning models, such as the GMM (Gaussian Mixture Model) [12]. In the classification stage, we seek to associate ROIs with real-world objects by extracting and summarizing features in each ROI. These first two stages involve other widely discussed problems, such as image processing, image segmentation, and topics associated with object classification [13–15]. The third stage, also called tracking, consists of identifying the behavior of an object based on its motion over time in a series of images [10]. This technique is known as object tracking and it can save multiple operations to segment and detect the same object in a sequence of images. Some of the algorithms for object tracking are: background region tracking, object tracking, object feature based tracking, and deep learning-based tracking [12]. In [16], a process to extract information based on object motion is addressed, using a sequence of images and tracking their trajectory. The trajectory of an object is usually obtained by tracking the (x, y) coordinates of its centroid frame by frame [3]. In [17], although no cameras are used, a pickpocket detection process is established, based on the anomalies identified in the routes they follow in public transport service networks, and for which information from electronic ticketing systems was used. In [18], they use the adaptive Optical Flow algorithm, which allows tracking identified objects among images in a video. Object tracking is useful to solve the personal re-identification problem [10], which consists of identifying the same person in a video sequence or from one camera to another, recognizing them instead of generating the entire identification process. The fourth stage seeks to automatically generate understanding of the images, that is, interpret what is happening in them, integrating the results of the previous stages and determining whether or not a situation of interest exists. The fifth stage, person identification, is usually treated separately, as in the case of facial identification and anti-spoofing [19], where people represent the main object of interest to achieve understanding. In [20], it is highlighted that one of the main objectives of intelligent video surveillance is the detection of people in order to try to understand, learn and recognize their abnormal behavior. An abnormal or different behavior in people can lead us to find potential threats. Below is a comparison table (Table 2), with the strategies used in different investigations to solve the stages of Detection, classification and tracking of objects in video analysis.

3.2 Related Works

In [27], they propose a method to detect abnormal crowd flows with an algorithm based on the mean displacement tracking model, also considering the behavior of the crowds and the groups that form in them. They also reference other models for pedestrian detection and tracking such as the Gaussian mixture tracking model and the artificial neural network model. In [28], they propose a Gaussian kernel-based integration model (GKIM), for the detection and localization of anomalous entities in pedestrian flows, focused on safeguarding pedestrian safety using video surveillance. An anomaly is considered to be anything or situation that has an unexpected movement (behavior) (pedestrians running, too many people, few people) and it is established that the problem of detecting anomalous entities in pedestrian flow is an important challenge due to the heterogeneity of the entities, the temporal variations and the changes in viewing angles. In [29], they implement a vehicle counting system using YOLOv5 for detection and SORT as a tracking system, adding an image cropping strategy to improve time and accuracy in the results. In [30, 31], they present a technique to detect anomalies in pedestrian areas using Deep Learning, classifying as anomalies everything that moves through the area, other than pedestrians, such as cars, skates, jeeps, etc. For this process, they rely on neural network models such as Mask-RCNN and DenseNet, or on Histogram of Optical Flow (HOOF). In [32], they introduce a method to detect abnormal events in crowds, focused on identifying situations of violent events or natural disasters. They use Optical Flow to obtain motion vectors and with them they train a convolutional neural network to classify whether it is an abnormal event. In [33], we present a self-learning system to count pedestrians in a specific region of interest (ROI), based on the characteristics of pedestrians in the scene and focusing especially on those that show little movement or are static. In [34] a passenger detection system is proposed for a transportation service using video surveillance and Deep Learning without the need to use GPUs since they only focus on detecting the person's head and recognizing their direction to determine whether they are entering or leaving the vehicle. In [35] a study on pedestrian counting with different known models is carried out and a combination of YOLOv3 with Deep SORT is tested, concluding that occlusion is the main source of error in these systems but that other factors such as speed, quantity and direction of pedestrian flow also influence. In [36], a system is presented to count pedestrians and determine their behavior when crossing a street, as well as calculate their waiting times. In this way, a traffic light can be configured or programmed to work more efficiently. In [37] a method is presented to detect, track and count pedestrians in real time, based on infrared images. In [38] a system for counting people grouped in different densities, with high occlusion, different background and different orientations of the video camera is presented.

3.3 Detection of People with YOLO

YOLO (You Only Look Once) is an image segmentation and object detection model that uses deep learning and computer vision, it was developed by Joseph Redmon and Ali Farhadi at the University of Washington in 2015 [24] and has been widely used to date due to its great performance in terms of speed and accuracy as well as being easily adaptable to different hardware platforms, from edge devices to cloud APIs [25].

Table 2. Technologies used to solve the different stages of video analysis.

	Detection and classification	Tracking	Data Processing	Storage
[3]	Simplification of Background cut	Geometrical analysis		
[18]	Viola-Jones	Optical Flow	Quad-core 2.3 GHz	
[11]	Color-Shape Based Object detection	NA	Video Quality Enhancement (CLAHE)	
[21]		Association Analysis Model		Abnormal behavior database
[2]	Neural Network		Jetson TX2	Datasets
[22]	AdaBoost		Raspberry Pi	Storage Device
[23]	YOLOv8		CUDA: 0 Tesla T4	SSHD
[9]	GMM, GMG y ABL	TFRE		
[6]	CNN, MoG	KCF		Tube database
[7]	MoG in OpenCV	Prediction algorithm	i7-4770 CPU @3.4 GHz	GRAM datasets

YOLO uses a simple neural network to find objects in an image, delimit them, and establish the probability of which class they belong to. Each predicted object is enclosed in a bounding box called a region of interest or ROI. To achieve this, the system divides an image into $S \times S$ cells and each cell predicts B bounding boxes with a confidence score for each. The confidence score indicates how confident the model is that the bounding box contains a given object and also how accurate the box is. In this way, two probabilities are generated: one to determine whether the bounding box contains an object and the other is to determine the class of the object being predicted. All of this can be reduced to a vector and as follows:

$$y^T = \left[p,\, p_1,\, p_2,\, p_3 \ldots p_k,\, x,\, y,\, w,\, h \right]$$

where p_i represents the probability that each of the k classes is predicted, for example, $p_1 = Person$, $p_2 = Cat$ and $p_3 = Dog$, and so on. On the other hand, p is the probability that the detected object fits well in that box with respect to an ideal box and is called confidence. And so, the size of the vector y will be $R = 4 + 1 + k$. Now, as the image was divided into $S \times S$ cells, at least each of them will have its own vector y, and if we consider that within the same cell B objects can be detected, then we will have Q elements for each cell, where $Q = B \times R$ [26], as illustrated in Fig. 1.

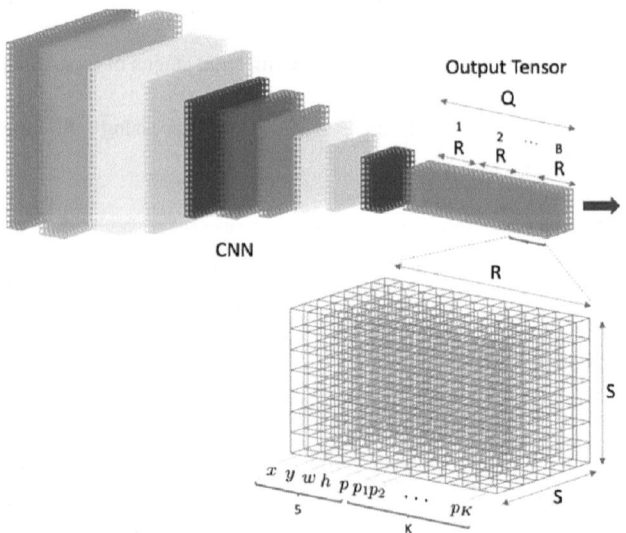

Fig. 1. YOLO uses a simple neural network dividing the image into several cells and each cell generates its own vector y which can contain several objects (Image taken from [26]).

4   Methodology

There is a historical database (see Fig. 2), which is processed at the start of each day (e.g., at midnight of each day), extracting the average daily pedestrian flow over fixed time periods, e.g., half-hour periods would generate a list of 48 periods, each of which represents the expected pedestrian flow for that time period. In addition, the standard deviation for each period is also calculated. Each period has the following attributes: *periodNumber, periodStartTime, periodEndTime, periodAverage* and *periodStandard-Deviation*. For database management, we work with the Python pymongo library to generate queries to the database built with MongoDb. For video capture, we used Tapo C200, C325 and C420 smart IP cameras with the Python cv2 library. For detection, we used the pre-trained version of YOLOv8n, configured to detect only people starting from a threshold of 60% and a very small IoU (intersection over union) to avoid duplications (0.01). For tracking, we tested the Byte Track x and BoT-SORT trackers, with the latter yielding the best results. Thus, when a person enters a scene, YOLO assigns them a one-time ID when they are detected as a person, at which point the event is saved in a MongoDB database. The event type, which is a scene entry, the date, time, and assigned ID are saved. The Bot-SORT tracker then predicts the person's movements in the scene, preventing it from being detected as a new person. Once this person leaves the scene, a new exit event is stored in the MongoDB database and the person is counted as part of the pedestrian flow. Finally, the real-time count of the pedestrian flow is compared to previously loaded historical statistics and assigned a degree of abnormality, ranging from zero to ten, where zero is considered a normal flow close to the average and ten as an abnormal flow. A threshold will determine the degree of abnormality at which the system will raise an alert, i.e., when the pedestrian flow tends to be abnormal. The way to assign the degree of abnormality is as follows:

$$GA = |(CF - HA)/(SD)(10)|$$

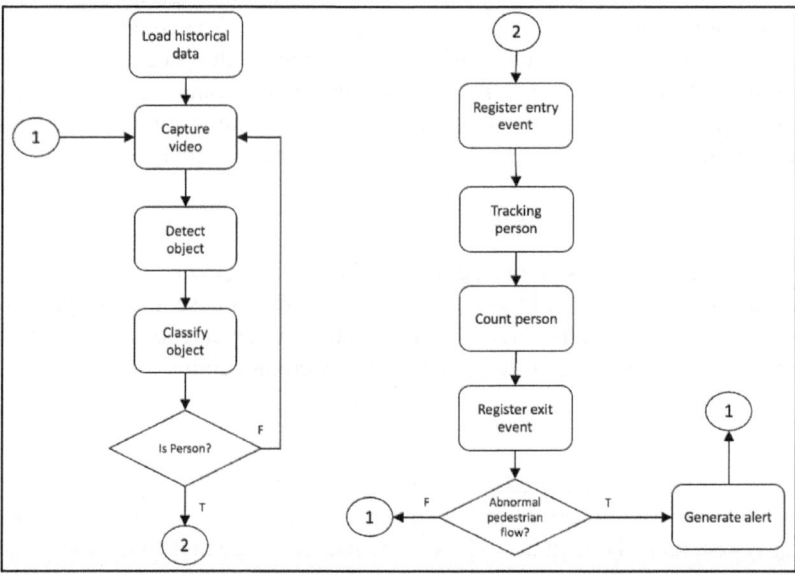

Fig. 2. Methodology for detecting abnormal pedestrian flows.

Where *GA* is the degree of abnormality of the pedestrian flow, *CF* is the current pedestrian flow that is being detected in real time, *HA* is the historical pedestrian flow calculated for pre-established time periods from a database that is fed back day by day and *SD* is the standard deviation of the historical pedestrian flow, also calculated from historical records. For this research, days were considered homogeneous and it will be pending for future work to consider that in many cases days could vary considerably in their pedestrian flow for example, working days and non-working days, possible quarantines, periodic social events, etc. On the other hand, at the end of a time period, for example every half hour, if an alert is not triggered for high pedestrian flow, the system will check that the pedestrian flow is not abnormally low, which will also generate an alert for low pedestrian flow, since excessive pedestrian flow can be as abnormal as low pedestrian flow based on historical pedestrian flows.

5 Results

To evaluate our proposal, we separated the research into two parts: one is responsible for counting pedestrian flow in real time and the other part is the one that generates historical statistics to determine when to launch an alert for abnormal pedestrian flow. Using videos with different previously recorded pedestrian flows, we first evaluated the pre-trained versions of YOLOv8n, YOLOv9m and YOLOv10n, executed on the same computer (M1 Pro with 16 Gb of RAM and MacOS Sonoma 14.6), obtaining significantly different execution times, with version 8n being the best evaluated while version 9m doubled the response time and version 10n also did not give good results. For this reason, we finally used the YOLOv8n version for people detection and as part of the solution for pedestrian flow counting, obtaining the following results through a confusion matrix. For the Precision indicator, which considers true positives (*TP*) and false positives (*FP*), where true positives are those people who are really part of the pedestrian flow and who received an identifier at least once from the pedestrian flow counting system, while false positives (*FP*) are those people who received a new identifier even though they already had one previously assigned, so the system counts them repeatedly.

$$Precision = TP/(TP + FP)$$

Another indicator that is often calculated is the sensitivity (*Recall*), which takes into account true positives (*TP*) and false negatives (*FN*). In our case, false negatives are all those people who, being really part of the pedestrian flow, never received an identification from the pedestrian counting system, that is, they were not counted.

$$Recall = TP/(TP + FN)$$

Specificity was also calculated, which includes true negatives (*TN*) and false positives (FP). In our case, true negatives are people who received an *ID* for being part of the counting system and who kept it for the entire time they were in the video sequence, that is, they were not counted repeatedly. This indicator shows the robustness of the tracking system being used. On the other hand, false positives count the *IDs* assigned repeatedly to people who are part of the pedestrian flow, that is, the same person can be counted several times.

$$Specificity = TN/(TN + FP)$$

Finally, another indicator called F_1, combines precision and sensitivity into a single metric to get a more complete view of model performance:

$$F_1 = 2(Precision)(Recall)/(Precision + Recall)$$

Tests were performed on several videos, with different conditions and times of pedestrian flows to calculate the confusion matrix and the following results were averaged: *Precision* = 90%, *Recall* = 84%, *Specificity* = 62.5% and F_1 = 87%. Below are some screenshots of the system running in the testing phase.

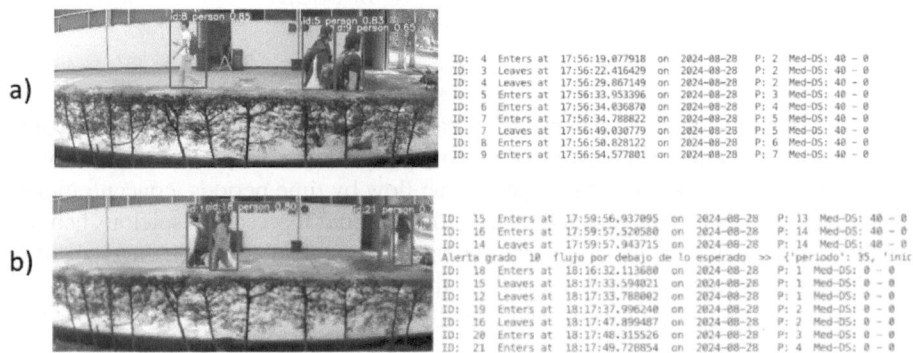

Fig. 3. Two screenshots of the pedestrian flow counter execution, a) on the left, YOLOv8n running and on the right, data thrown during the execution. b) An alert is activated due to low pedestrian flow

The data shown in Fig. 3, is the *ID* assigned by the YOLOv8n.pt algorithm, then the time and date when the person enters the scene is specified, that is, when an *ID* is assigned, the *P* column indicates the people counted at that time, this counter is reset in each period, which is configured for 30 min. Below are the values of the mean and standard deviation of the pedestrian flow for that period, this is obtained each day from a historical database made in MongoDB, in our case it is configured to load the updated data at the beginning of each day (approximately zero hours). The database is updated in real time with each event that is generated in the scene, whether it is the entry or exit of a person. Finally, the system generates alert notices when the pedestrian flow values exceed the expected values or if at the end of each period the pedestrian flow is classified as scarce. Table 3 compares the accuracy of pedestrian counting with some recent research.

Future Work. Thanks to the construction of the historical database, new indicators can still be inferred, for example, statistics can be generated on how long it is reasonable for a person to remain in the place depending on the context of the scene (place, date and time);

Table 3. Comparing pedestrian counting accuracy with some recent research.

	Precision	Detection	Tracking	Real-Time
[34]	99%	Tiny-YOLOv3	KCF	✅
[35]	83%−	YOLOv3	Deep SORT	🚫
[36]	72%	Haar-cascade	SORT	✅
[37]	82.11%	Haar-like features	Kalman Filter	✅
[38]	86.90%	HOG	Motion Trajectories	🚫
Our proposal	90%+	YOLOv8 PT	BoT-SORT	✅

if a person remains longer than expected, it can be considered as anomalous behavior. It is also possible to know at all times how many people there are per frame in the video, in the same way, beyond measuring pedestrian flow by time periods, concentrations of people can be known at unexpected times and that could represent suspicious behavior. Another aspect is that the same system begins to automatically catalog the types of day to make better estimates of pedestrian flow, that is, distinguish between work days and non-work days, or sporadic events associated with a date or time.

6   Conclusions

A methodology was presented and an application was developed for the detection of abnormal pedestrian flows using a pre-trained neural network (Ultralytics YOLOv8) that, based on historical statistics of the environment, issues alerts if it detects a pedestrian flow above or below the expected values, considering confidence thresholds based on the standard deviation. The system validation was carried out in environments with different pedestrian flows and a counting/tracking strategy based on the assignment of *IDs* by the pre-trained network tracking system was validated, obtaining the following indicators: *Precision* = 90%, *Recall* = 84%, *Specificity* = 62% and F_1 = 87%.

Acknowledgments. We would like to acknowledge a group of students from the Instituto Tecnológico de Cd. Hidalgo, who participated in the Delfin 2024 summer program and were collaborating in part of this project, they are: Alexis Baca Cruz, Edgar Brandom Carrillo Vaca and Juan David Mora Sandoval.

Disclosure of Interests. The authors would like to thank the Tecnológico Nacional de México for the financial support provided through project 19476.24-P.

References

1. Sistema Nacional de Seguridad Pública, Norma Técnica para estandarizar las características técnicas y de interoperabilidad de los sistemas de vídeo vigilancia para la seguridad pública. Centro Nacional de Información, México (2018)
2. Edwin, J., Greeshma, M., Mithun-Haridas, T.P., Supriya, M.H.: Face recognition based surveillance system using FaceNet and MTCNN on Jetson TX2. In: ICACCS, pp. 608–613 (2019)
3. Pritch, Y., Ratovitch, S., Hendel, A., Peleg, S.: Clustered synopsis of surveillance video. In: 2009 Sixth IEEE International Conference on Advanced Video and Signal Based Surveillance, Genova, Italy (2009)
4. Ahmed, A., Dogra, D.P., Kar, S., Roy, P.P.: Trajectory-based surveillance analysis: a survey. IEEE Trans. Circuits Syst. Video Technol. **29**(7), 1985–1997 (2019)
5. Ghatak, S., Rup, S.: Single camera surveillance video synopsis: a review and taxonomy. In: 2019 International Conference on Information Technology (ICIT), Bhubaneswar, India (2019)
6. Ahmed, S.A., et al.: Query-based video synopsis for intelligent traffic monitoring applications. EEE Trans. Intell. Transp. Syst. **21**(8), 3457–3468 (2020)
7. Ratnarajah, A.J., Goonetilleke, S., Tissera, D., Balagopalan, K., Rodrigo, R.: Moving object based collision-free video synopsis. In: IEEE International Conference on Systems, Man, and Cybernetics (SMC), Miyazaki, Japan (2018)
8. INEGI: Encuesta Nacional de Victimización y Percepción sobre Seguridad Pública (ENVIPE). Subsistema de Información de Gobierno, Seguridad Pública e Impartición de Justicia (2022). https://www.inegi.org.mx/temas/incidencia/. Accessed 01 May 2023
9. Kong, L., Dai, R.: Object-detection-based video compression for wireless surveillance systems. IEEE Multimedia **24**(2), 76–85 (2017)
10. Zhang, Q., Sun, H., Wu, X., Zhong, H.: Edge Video Analytics for Public Safety: A Review. Proc. IEEE **107**(8), 1675 (2019)
11. Xiao, J., Li, S., Xu, Q.: Video-based evidence analysis and extraction in digital forensic investigation. IEEE Access **7**, 55432–55442 (2019)
12. Ye, Y., Ci, S., Katsaggelos, A.K., Liu, Y., Qian, Y.: Wireless video surveillance: a survey. IEEE Access **1**, 646–660 (2013)
13. Michal, G., Andzej, M., Piotr, G., Mikolaj, L.: Automated detection of firearms and knives in a CCTV image. Sensors **16**(1), 47 (2016)
14. Surajit Saikia, E., Fidalgo, E.A., Fernández-Robles, L.: Object detection for crime scene evidence analysis using deep learning. In: Battiato, S., Gallo, G., Schettini, R., Stanco, F. (eds.) ICIAP 2017. LNCS, vol. 10485, pp. 14–24. Springer, Cham (2017). https://doi.org/10.1007/978-3-319-68548-9_2
15. Redmon, J., Farhadi, A.: axXiv Forum (2018). https://arxiv.org/abs/1804.02767. Accessed 28 Mar 2023
16. Teddy, K.: A survey on behavior analysis in video surveillance for homeland security applications. In: IEEE Applied Imagery Pattern Recognition Workshop, Washington, DC, USA (2008)
17. Bowen, D., Chuanren, L., Wenjun, Z., Zhenshan, H., Hui, X.: Detecting pickpocket suspects from large-scale public transit records. IEEE Trans. Knowl. Data Eng. **31**(3), 465–478 (2019)
18. Elrefaei, L.A., Alharthi, A., Alamoudi, H., Almutairi, S., Al-rammah, F.: Real-time face detection and tracking on mobile phones for criminal detection, Abha, Saudi Arabia. IEEE Xplore (2017)
19. Yu, Z., Qin, Y., Li, X., Zhao, C., Lei, Z., Zhao, G.: Deep learning for face anti-spoofing: a survey. IEEE Trans. Pattern Anal. Mach. Intell. **45**(5), 5609–5631 (2023)

20. Magadan-Salazar, A.: Sistemas de videovigilancia inteligentes. Komputer Sapiens, vol. II, no. X, pp. 22–26 (2018)
21. Shao, Z., Cai, J., Wang, Z.: Smart monitoring cameras driven intelligent processing to big surveillance video data. IEEE Trans. Big Data **4**(1), 105–116 (2018)
22. Suleman, K., Hammad, J.M., Ehtasham, A., Syed, S., Syed, U.A.: Facial recognition using convolutional neural networks and implementation on smart glasses. In: International Conference on Information Science and Communication Technology (ICISCT), Karachi, Pakistan (2019)
23. Pérez-García, G., Escobar-Gómez, E., Sarmiento-Torres, J., Juárez-Ruiz, I., Flores-García, A.J.: Intelligent weapon detection system using computer vision. Tecnología y Ciencia Aplicada **6**, 26–31 (2023)
24. Redmon, J., Divvala, S., Girshick, R., Farhadi, A.: You only look once: unified, real-time object detection. In: IEEE Conference on Computer Vision and Pattern Recognition (CVPR), Las Vegas, NV, USA (2016)
25. Jocher, G., Chaurasia, A., Munawar, M.R.: Ultralytics YOLO Docs (2024). https://docs.ult ralytics.com. Accessed 08 Aug 2024
26. Mery, D.: 18 Visión por computador: object detection – YOLO. YouTube (2021). https:// www.youtube.com/watch?v=-VcyIt0p7bA&list=PLilWJnCHHGl2Iog1Tusf62T3cq7Vz 3UoT&index=18. Accessed 21 July 2024
27. Ma, J., Song, W.: Automatic clustering method of abnormal crowd flow pattern detection. Procedia Eng. **62**, 509–518 (2013). ISSN: 1877-7058
28. Ullah, H., Altamimi, A.B., Uzair, M., Ullah, M.: Anomalous entities detection and localization in pedestrian flows. Neurocomputing **290**, 74–86 (2018). ISSN: 0925-2312
29. Valencia, D., Muñoz, E., Muñoz-Añasco, M.: Impact of the preprocessing stage on the performance of offline automatic vehicle counting using YOLO. IEEE Lat. Am. Trans. **22**, 723–732 (2024)
30. Pustokhina, I.V., Pustokhin, D.A., Vaiyapuri, T., Gupta, D., Kumar, S., Shankar, K.: An automated deep learning based anomaly detection in pedestrian walkways for vulnerable road users safety. Saf. Sci. **142** (2021). ISSN: 0925-7535
31. Ramalingam, V.V., Patni, S., Mohan, S.G.: Towards detection of abnormal event and reporting for pedestrian video surveillance. Int. J. Health Sci. **6**, 4440–4455 (2022)
32. Direkoglu, C.: Abnormal crowd behavior detection using motion information images and convolutional neural networks. IEEE Access **8**, 80408–80416 (2020)
33. Huang, L., Liu, C., Li, J.: An efficient self-learning people counting system. In: The First Asian Conference on Pattern Recognition, Beijing (2011)
34. Sohn, M.K., Lee, S.H., Kim, H.: Development of a real-time automatic passenger counting system using head detection based on deep learning. J. Inf. Process. Syst. **18**, 428–442 (2022)
35. Meli, W., Lacy, F., Ismail, Y.: Video-based automated pedestrians counting algorithms for smart cities. Int. J. Comput. Digit. Syst. **6**, 1065–1078 (2020)
36. Wickramasinghe, K.S., Ganegoda, G.U.: Pedestrian detection, tracking, counting, waiting time calculation and trajectory detection for pedestrian crossings traffic light systems. In: 2020 20th International Conference on Advances in ICT for Emerging Regions (ICTer), Colombo, Sri Lanka (2020)
37. Shahzad, A.R., Jalal, A.: A smart surveillance system for pedestrian tracking and counting using template matching. In: 2021 International Conference on Robotics and Automation in Industry (ICRAI), Rawalpindi, Pakistan (2021)
38. Pervaiz, M., Ghadi, Y.Y., Gochoo, M., Jalal, A., Kamal, S., Kim, D.S.: A smart surveillance system for people counting and tracking using particle flow and modified SOM. Sustainability **13**, 1–20 (2021)

Computational Time Reduction in the Induction of Convolutional Decision Trees

Adriana-Laura López-Lobato$^{(\boxtimes)}$ (ID), Héctor-Gabriel Acosta-Mesa(ID), and Efrén Mezura-Montes(ID)

Artificial Intelligence Research Institute, University of Veracruz, Campus Sur, Calle Paseo Lote II, Sección Segunda No. 112, Nuevo Xalapa, 91097 Xalapa-Enríquez, Veracruz, Mexico
adrilau17@gmail.com, {heacosta,emezura}@uv.mx
https://www.uv.mx/iiia

Abstract. Convolutional decision trees (CDTs) are machine learning models employed as explicable methods for image segmentation. This is because their graphical structure makes it relatively straightforward to interpret how the tree successively divides the image pixels into two classes, distinguishing between objects of interest and the image's background. Several techniques have been proposed for the induction of CDTs. However, these techniques often require significant computational time and memory, with some requiring days to complete the induction process. This study proposes two techniques for selecting a representative sample of pixels from an image for the model's training process: raw selection and median selection. These techniques aim to reduce the computational cost of inducing a CDT while maintaining or improving the resulting segmentation's precision, measured by the F1-score. The proposed techniques were evaluated using the retinal vessel segmentation database and the SHADE-CDT method to induce a CDT, demonstrating a 70% reduction in processing time.

Keywords: Image segmentation · Convolutional Decision Trees · SHADE

1 Introduction

Semantic segmentation is a process of interest in computer vision that consists of differentiating an object from the background of the image by assigning a label to the pixels of the image. While several methods have been proposed to solve the image segmentation problem [4,7], this work takes a novel approach by studying the relatively unexplored Convolutional Decision Trees (CDTs) [3]. These machine learning models are employed as interpretable methods for image segmentation due to their graphical structure, which enables the interpretation of how the tree successively divides the pixels into two classes, assigning a label

L. Martínez-Villaseñor et al. (Eds.): MICAI 2024 Workshops, LNAI 15465, pp. 99–111, 2025.
https://doi.org/10.1007/978-3-031-83882-8_10

(1 or 0) to each pixel of an image and distinguishing between objects of interest and the background of the image, thereby performing the segmentation task [5,6].

Several techniques have been proposed to address the optimization challenges inherent in the CDT induction process. The original approach focuses on maximizing the information gain function at each tree node through an analytical optimization process [3]. Other methods leverage the capabilities of the Differential Evolution (DE) algorithm, a robust and versatile optimization technique, to conduct both global and local searches [2,5]. These searches aim to identify convolutional kernels for CDT nodes that enhance the F1-score, a key metric in image segmentation performance. The SHADE algorithm has also been employed to induce CDTs locally and globally [6].

However, these techniques require a considerable amount of computational time and memory for the induction process, since the classification of each image pixel in the training set is needed to evaluate the fitness function in the differential evolution process. This evaluation must be performed for each CDT proposed as an individual of the DE algorithm and for each generation.

This work introduces two techniques for selecting a representative sample of pixels from an image for the model's training process: raw selection and median selection. This article compares these techniques with the SHADE-CDT induction procedure, proposed in [6], using the segmentation results of the "Retinal Vessel Segmentation" dataset. This database was selected since the full dataset can induce a CDT in a reasonable time with the SHADE-CDT process and can be compared with the results of the proposed computational cost reduction techniques. The results of this study demonstrate that the proposed techniques significantly reduce the computational time required for CDT induction while maintaining or enhancing the precision of the resulting segmentation. The potential benefits of these techniques suggest a promising avenue for further research in computational intelligence and image processing.

The remaining paper is structured into three sections. In Sect. 2, a brief description of the DE algorithm and the SHADE algorithm is shown. This section also presents a concise overview of the SHADE-CDT induction procedure. Section 3 is dedicated to the experiments and the results obtained. Finally, Sect. 4 offers detailed conclusions and future work.

2   Methods

This section presents a theoretical overview of the Differential Evolution (DE) Algorithm and its variant, SHADE. Additionally, it provides a description of Convolutional Decision Trees (CDTs) and the methodology SHADE-CDT employed to induce CDTs with a global strategy.

2.1   Differential Evolution Algorithm and SHADE

The Differential Evolution (DE) Algorithm is a metaheuristic research strategy that addresses optimization problems [8]. The DE algorithm is therefore employed when the optimization of a given fitness function is required. In order to optimize this fitness function with the DE Algorithm, a set of candidate solutions (a population of individuals) is created, considering elements on the function's domain. The fitness measure of each individual is obtained, and this information is used to select some candidates to seed the next generation. This is done by applying two variation operators, crossover and mutation.

The crossover operator is applied to two or more selected individuals, denoted as *parents*, resulting in the generation of one or more new solutions, denoted as *children*. The mutation operator is employed to generate a new candidate from a single existing candidate. To create new candidates, called *offspring*, with sufficient diversity to facilitate the search, two parameter values must be input for these variation operators: the scaling factor, denoted as F, and a crossover rate, represented by CR. Using a selection operator, which forces fit individuals to advance, the offspring's fitness is compared to the preceding generation's fitness to secure a position within the subsequent generation.

The generation of a new population from a previous one is repeated several times until a stopping criterion is met. The best solution to the problem found with the DE Algorithm is identified by considering the individual with the best fit within the population obtained in the final generation.

To run the Differential Evolution process, the user must input several parameter values: the crossover rate (CR), the scale factor (F), the population size (NP), and the number of generations (NG). These values are problem-depend and directly impact the performance of the DE algorithm [1]. Therefore, it is necessary to tune them to obtain good results. For these reasons, several self-adaptive techniques have been investigated with a view to adjusting these parameters. These include JADE [11], SHADE [9], and L-SHADE [10].

This paper considers the SHADE algorithm. The Success History-based Adaptive Differential Evolution (SHADE) algorithm employs a historical memory, of size H, with successful parameters CR and F, to adjust their values on each iteration of the search process. Further details on the SHADE algorithm can be found in [9].

2.2   Convolutional Decision Trees

The Convolutional Decision Trees (CDTs) are multivariate decision trees, proposed for the first time in [3], where each condition is given by a convolutional kernel, see Fig. 1. The convolutional kernels are of interest in image processing and computer vision since a convolution process on an image results in a filtered version of the image, where specific features are extracted.

Several techniques have been proposed to solve the optimization problems inherent to the CDT induction process [2,3,5,6], however, in this work, the SHADE-CDT method to induce a CDT was selected to analyze the two computational cost reduction techniques proposed. This method is a global search strategy proposed in [6], where the SHADE algorithm is used to induce a CDT of a given depth d with kernels of size s.

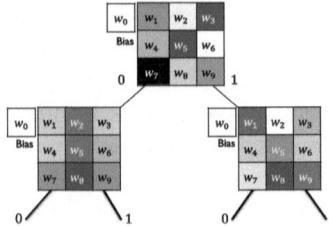

Fig. 1. Example of a CDT of depth 2 and kernel size 3.

This process assigns a label to each pixel with a CDT, by passing the pixel information (encoded as a vector with the values of the pixels in the neighborhood of size $s^2 + 1$ surrounding it with a bias value of 1) through the corresponding kernels (encoded as real-valued vectors of size $s^2 + 1$ with the weights of the convolutional kernel associated with the corresponding internal node of the CDT and a value for the bias) until it reaches a leaf node, see Fig. 2.

The dot product between the pixel's encoded vector and the corresponding kernel's encoded vector passed through an activation function that returns a label 0 or 1, and this label indicates the subsequent node to which the instance will proceed. The kernel on the left branch is associated with label 0, while the kernel on the right branch is associated with label 1. The label assigned to the instance is obtained using the corresponding kernel in the leaf node, see Fig. 2.

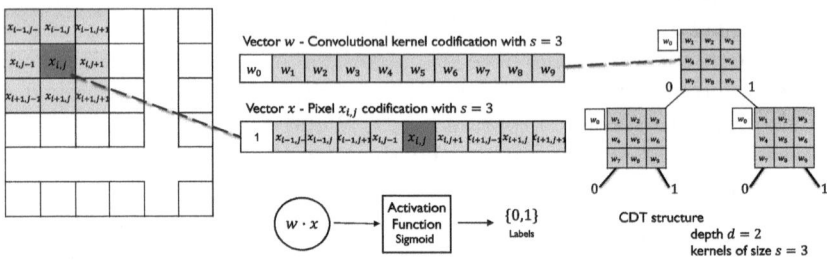

Fig. 2. Codification of a convolutional kernel and codification of a pixel-associated instance, and how these elements are processed in a CDT with kernel size $s = 3$.

To apply the SHADE algorithm, the F1-score metric is employed as the fitness function value. The F1-score is a metric used to evaluate the accuracy of a model's label assignment by comparing the predicted labels with the actual labels of the instances. The individuals in the population are vectors of size $(s^2 + 1)(2^d - 1)$ composed of random values drawn from -255 to 255. The values in the vector correspond to internal convolution kernels and their corresponding biases for the tree.

In this method, the population size (NP), the number of generations (NG), and the memory size (H) are user-defined parameters. Further details on the SHADE-CDT method can be found in [6].

2.3   Proposed Methods

The techniques proposed in this paper aim to reduce the computational time required to generate a CDT by obtaining a set of pixels from the original training set images that are representative of the features associated with each class. In this way, the proposals reduce the computational time required for the training process while maintaining the results of a CDT generated with the information from the entire training set.

The following section describes the proposed techniques in detail.

Computational Cost Reduction Techniques

The techniques proposed in this section involve the selection of a sample of representative pixels from each of the two classes to be segmented in the images. These techniques are distinguished by the selection process made: row or median. In the process of row selection, the sample of pixels used is obtained by randomly selecting pixels from each class. In the process of median selection, a partition of the pixels is obtained to calculate the median vector with the representative information of each subset. Recall that a partition is a division of a set into a family of subsets that are mutually disjoint and jointly exhaustive; that is, each element of the original set is present only in one of the subsets, and all the subsets together contain all the members of the original set.

For both techniques, the following steps are performed on each image of the training set, taking into account step 3.1 for the raw selection technique or step 3.2 for the median selection technique:

1. The labels of the pixels in the image are obtained and divided, forming two sets of indices, I_0 and I_1, for each class of pixels, class 0 and class 1, respectively, see Fig. 3.

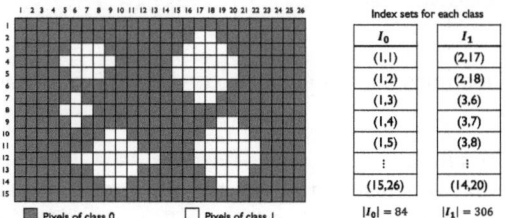

Fig. 3. Class labels of an image and their division into sets I_0 and I_1.

2. To obtain the *number of pixels* or the *number of subsets* for the proposed methodologies, a proportion $\mathcal{P}$ previously established by the user is needed. This value indicates the proportion of the total image information sampled for

the induction process. For example, if the user sets a proportion $\mathcal{P} = 0.1$, the process will seek to obtain a representative sample of a size equivalent to 10% of the number of pixels in the analyzed image, maintaining the proportion between the classes.

In this step the *floor* function, in Eq. (1), is employed. This function returns as a result the greatest integer less than or equal to the real number x entered as an input.

$$\lfloor x \rfloor = min\{n \in Z : n \geq x\} \tag{1}$$

The number of pixels or subsets is determined by evaluating the product of the set size, denoted as $|I_i|$ ($i = 0, 1$), and the proportion specified by the user in the floor function, as illustrated in Eqs. (2).

$$P_0 = \lfloor |I_0| * \mathcal{P} \rfloor \text{ and } P_1 = \lfloor |I_1| * \mathcal{P} \rfloor. \tag{2}$$

In the illustrative example represented in Fig. 3, the following calculations are conducted when the proportion $\mathcal{P}$ is set to the value 0.1.

$$P_0 = \lfloor 84 * 0.1 \rfloor = \lfloor 8.4 \rfloor = 8 \text{ and } P_1 = \lfloor 306 * 0.1 \rfloor = \lfloor 30.6 \rfloor = 30. \tag{3}$$

In the case of the class 0, eight pixels or subsets of the dataset are taken into consideration, whereas for the class 1, there are thirty.

3. To conform the representative vectors of the images in the training dataset, the following process are followed on each methodology:
    3.1 **Raw selection.**
    3.1.1 A random selection of P_0 pixels of class 0, and a random selection of P_1 pixels of class 1 is made.
    3.1.2 The encodings of the instances associated with these pixels are obtained, as illustrated in Fig. 4.

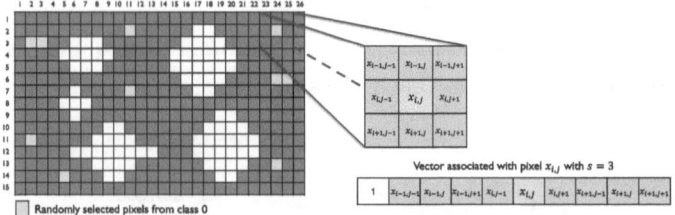

Fig. 4. Encoding an instance that is associated with a pixel of the randomly selected pixels from class 0.

3.2 **Median selection.**
3.2.1 The *size of the subsets*, that is, the number of pixels that each subset has, is determined by evaluating the floor function of the ratio between the number of indices in the set and the corresponding number of subsets calculated in the preceding step, as illustrated in Eqs. (4).

$$PS_0 = \left\lfloor \frac{|I_0|}{P_0} \right\rfloor \text{ and } PS_1 = \left\lfloor \frac{|I_1|}{P_1} \right\rfloor \tag{4}$$

In the example, the following calculations are performed

$$PS_0 = \left\lfloor \frac{84}{8} \right\rfloor = \lfloor 10.5 \rfloor = 10 \text{ and } PS_1 = \left\lfloor \frac{306}{30} \right\rfloor = \lfloor 10.2 \rfloor = 10. \quad (5)$$

The results indicate that each subset of the dataset I_0 must consist of ten pixels, and similarly, each subset of the dataset I_1 must also comprise ten pixels.

3.2.2 A subset of the indices in I_0 and I_1 is obtained by randomly selecting PS_0 indices from the set I_0 and PS_1 indices from the set I_1. The selected indices are then discarded, and this process is repeated until the requisite number of subsets has been reached, see Fig. 5.

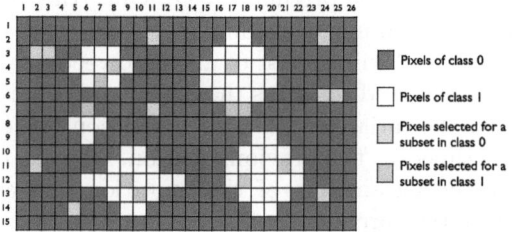

Fig. 5. Example of subsets with ten elements for each class.

3.2.3 The encodings of the instances associated with the pixels of the subset are obtained, as illustrated in Fig. 6, and the median value is calculated, coordinate by coordinate. This process is illustrated in Fig. 7.

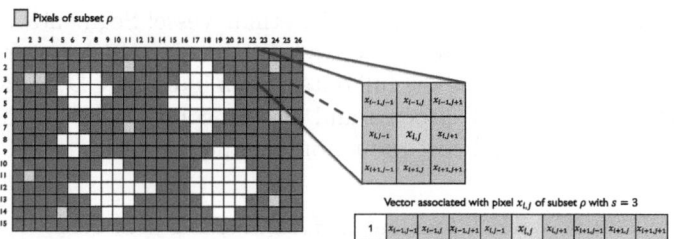

Fig. 6. Encoding an instance that is associated with a pixel of subset ρ.

3.2.4 The mean vectors are then assigned to the class corresponding to the subset from which they were created. Subsequently, the process generates P_0 elements of class 0 and P_1 of class 1.

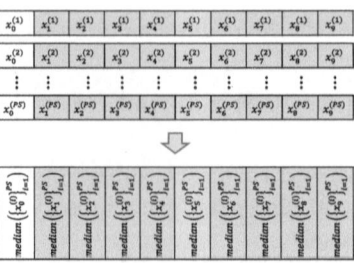

Fig. 7. Obtaining the representative vector for the analyzed subset.

As previously described, these processes yields $P_0 + P_1$ representative vectors of the analyzed image's information. The resulting vectors are consistent with the proportions observed in the original image. Once this process of obtaining representative vectors of each image is complete, the general training set for the induction process of a CDT will consist of all these vectors.

The experiments conducted to assess the efficacy of the proposed approaches were carried out using the variant of the CDT induction process with SHADE in the global way, as outlined in the previous section. The use of SHADE enables precise control over the parameters of the mutation and crossover operators in the differential evolution process.

The proposed strategies were implemented using the Matlab R2023b software on a PC with the OS Windows 11 Pro 23H2, 64 GB of RAM, and an AMD Ryzen 5 5600G processor operating at a speed of 3.90 GHz.

3 Experiments and Results

In this section, some results obtained by inducing a CDT with the SHADE-CDT process are reported using the database "Retinal Vessel Segmentation"[1], where 20 images are shown, with their respective segmentation masks or ground truth. The images were selected since the full dataset can induce a CDT in a reasonable time with the SHADE-CDT process and can be compared with the results of the proposed computational cost reduction techniques.

3.1 Experiments with Classical SHADE-CDT Process

Controlled experiments were carried out to calibrate the parameter values of the SHADE-CDT process with all the information in the training set images. The resolution of the images is originally 565 × 584, but for experimentation purposes, they are resized to 256 × 256. For these experiments, 70% of the images are considered for training and 30% for testing. The parameter settings of these experiments are presented in Table 1.

Table 2 shows the image segmentation results of the images in the test set, employing the SHADE-CDT method with all pixels from the training dataset

---

[1] https://drive.grand-challenge.org/DRIVE/.

Table 1. Parameter settings for the SHADE-CDT process.

Parameter	Value	Parameter	Values
Population size (NP)	100	Kernel sizes (k)	3, 5, 7, 9
Number of generations (NG)	200	Tree depth (d)	2, 3, 4
H	100		

for the CDT induction. The tree depth and kernel size were modified, while the training and test datasets were maintained between experiments to ensure consistent comparison. The highest F1-scores were achieved with a kernel size of 5 and a depth of 2. Consequently, these two values were maintained for subsequent experiments with the computational cost reduction techniques.

Table 2. Image segmentation results for the images in the test set using the SHADE-CDT method for CDT induction, with all pixels from the images in the training set. Notes: The F1-score and accuracy metrics were calculated using the actual and predicted labels of the images in the test set. Bold numbers indicate the best results by kernel size, and the best result is highlighted

Exp	Kernel size	Depth	Time(hr)	F1-score	Accuracy
1	3	2	1.2	0.59703	0.92557
2	3	3	1.47	0.60465	0.9272
3	**3**	**4**	**1.79**	**0.60763**	**0.92825**
4	**5**	**2**	**1.44**	**0.64068**	**0.93128**
5	5	3	1.77	0.63201	0.9321
6	5	4	2.04	0.59286	0.91506
7	7	2	1.72	0.58169	0.91254
8	**7**	**3**	**1.98**	**0.62121**	**0.92619**
9	7	4	2.52	0.52534	0.89585
10	**9**	**2**	**2.1**	**0.56278**	**0.90797**
11	9	3	2.69	0.50891	0.89439
12	9	4	3.27	0.46263	0.87092

3.2 Experiments with the Computational Cost Reduction Techniques in the SHADE-CDT Process

In the experiments conducted to evaluate the computational cost reduction techniques proposed in this paper, the conditions of the images were maintained consistent with those of the previous experiments: resize to 256×256, 70% of the images for training and 30% for testing, considering the exact same images from the previous experiments with full data. Also, a proportion $\mathcal{P}$ is needed to determine the amount of information used in the process. Controlled experiments were performed to calibrate this parameter, considering proportions of 10%, 20%, 30%, 40% and 50%, obtaining the best results with 30%, corresponding to $\mathcal{P} = 0.3$, so this value was maintained for the subsequent experiments.

Both techniques, raw and median selection, with the SHADE-CDT process were tested under these conditions on 10 independent executions. Also the SHADE-CDT induction process with all pixels from the training dataset were perform 10 times. The parameter settings of these experiments are presented in Table 3.

Table 3. Parameter settings for the SHADE-CDT process.

Parameter	Value	Parameter	Value
Population size (NP)	100	Kernel size (k)	5
Number of generations (NG)	200	Tree depth (d)	2
H	100	Proportion ($\mathcal{P}$)	0.3

Table 4 illustrates the outcomes of the image segmentation process for the images included in the test set. The mean times reported for each method demonstrate that the two techniques proposed in this paper reduce the computational time required for the learning process in the induction of the CDTs, with a notable reduction of 70%.

Table 4. Obtained F1-score results with the SHADE-CDT method for CDT induction with all the pixels of the images in the training set and with the two computational cost reduction techniques, raw selection and median selection.

Method SHADE-CDT	Min F1-score	Max F1-score	Mean F1-score ± St.D.	Mean Time
Full Training Data	0.6128	0.6426	0.6283 ± 0.0089	1.43 h
Raw Selection	0.6118	0.6500	0.6368 ± 0.0118	27.09 min
Median selection	0.6318	0.6502	0.6415 ± 0.0069	27.21 min

The Shapiro-Wilk statistical tests was performed to verify the normal distribution on the Accuracy and the F1-score results obtained in the 10 executions of each of the three methods. Then, the one-way ANOVA test was performed to determine whether any of the differences between the means are statistically significant. Since the p-value obtained with this test (0.0136) is less than the significance level $\alpha = 0.05$, it is concluded that the differences between some of the methods means are significant.

As ANOVA rejects the null hypothesis that all methods means are equal, the multiple comparisons test is used to determine which means are different from others. Figure 8 illustrates the test result, that confirms the ANOVA analysis. The blue bar shows the comparison interval for the first method mean (SHADE-CDT with Full Training Data), which does not overlap with the comparison intervals for the third method mean (SHADE-CDT with Median Selection), shown in red. So, the means for the first and third methods are significantly different from each other. The comparison interval for the mean of second method (SHADE-CDT with Raw Selection), shown in gray, overlaps with the comparison interval for the first method mean. Hence, the means for the first and second methods are not significantly different from each other.

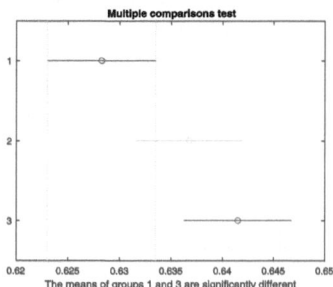

Fig. 8. Multiple comparisons test between the analized methods. Notes: 1 is the SHADE-CDT method with full training data, 2 is the SHADE-CDT method with Raw Selection, 3 is the SHADE-CDT method with Median Selection. (Color figure online)

It can be stated that the SHADE-CDT method with the raw selection technique produces comparable results to the same method when applied to the complete training data set. However, the computational time required for the learning process is reduced. Furthermore, the SHADE-CDT method with the median selection technique demonstrates superior performance in terms of segmentation results when compared to the full training data set process, while also requiring less computational time to induce the CDT.

4   Conclusions and Future Work

This work presents two techniques for reducing the computational time required to induce convolutional decision trees. The first technique is the raw selection, and the second is the median selection. These techniques were compared with the SHADE-CDT method proposed in [6]. The primary distinction between the proposed techniques and the SHADE-CDT method is that a representative sample of pixels from the training dataset is used to induce a CDT. Applying these techniques to the induction of CDTs results in a reduction in computational time, due to a reduction in the number of training instances. To illustrate, the time reduction in the experiments conducted in this paper was 70%, with the training dataset size reduced in proportion. It is also noteworthy that the proposed techniques maintain and even enhance the quality of the segmentation results, likely due to the reduction in the number of instances in the training set, which minimizes the risk of overfitting.

Further research could investigate the integration of these techniques into more sophisticated methods, such as semantic segmentation of color images with CDTs, since currently semantic segmentation methods with CDTs are limited to grayscale images due to the computational cost associated with this approach. Furthermore, it is essential to evaluate the efficacy of the proposed approach in comparison to alternative image segmentation methods, such as Convolutional Neural Networks. Moreover, additional metrics, such as the IoU, will be used to evaluate the results of the segmentation.

Acknowledgments. The first author is funded by the National Council of Humanities, Sciences and Technologies (CONAHCyT), through a postdoctoral fellowship at the Artificial Intelligence Research Institute of the University of Veracruz.

Disclosure of Interests. The authors have no competing interests to declare that are relevant to the content of this article.

References

1. Ahmad, M.F., Isa, N.A.M., Lim, W.H., Ang, K.M.: Differential evolution: a recent review based on state-of-the-art works. Alex. Eng. J. **61**(5), 3831–3872 (2022)
2. Barradas Palmeros, J.A., Mezura Montes, E., Acosta Mesa, H.G., Márquez Grajales, A., Rivera López, R.: Induction of convolutional decision trees with differential evolution for image segmentation. In: Proceedings: Congreso Mexicano de Inteligencia Artificial, p. 8 (2023)
3. Laptev, D., Buhmann, J.M.: Convolutional decision trees for feature learning and segmentation. In: German Conference on Pattern Recognition, pp. 95–106. Springer (2014)
4. Lateef, F., Ruichek, Y.: Survey on semantic segmentation using deep learning techniques. Neurocomputing **338**, 321–348 (2019)
5. López-Lobato, A.L., Acosta-Mesa, H.G., Mezura-Montes, E.: Blood cell image segmentation using convolutional decision trees and differential evolution. In: Advances in Computational Intelligence. MICAI 2023 International Workshops, pp. 315–325. Springer (2024)
6. López-Lobato, A.L., Acosta-Mesa, H.G., Mezura-Montes, E.: Induction of convolutional decision trees with success-history-based adaptive differential evolution for semantic segmentation. Math. Comput. Appl. **29**(4), 48 (2024)
7. Patil, D.D., Deore, S.G.: Medical image segmentation: a review. Int. J. Comput. Sci. Mob. Comput. **2**(1), 22–27 (2013)
8. Storn, R., Price, K.: Differential evolution-a simple and efficient heuristic for global optimization over continuous spaces. J. Glob. Optim. **11**, 341–359 (1997)
9. Tanabe, R., Fukunaga, A.: Success-history based parameter adaptation for differential evolution. In: 2013 IEEE Congress on Evolutionary Computation, pp. 71–78. IEEE (2013)

10. Tanabe, R., Fukunaga, A.S.: Improving the search performance of shade using linear population size reduction. In: 2014 IEEE Congress on Evolutionary Computation (CEC), pp. 1658–1665. IEEE (2014)
11. Zhang, J., Sanderson, A.C.: JADE: adaptive differential evolution with optional external archive. IEEE Trans. Evol. Comput. **13**(5), 945–958 (2009)

Bean Landraces Color Identification Through Image Analysis and Gaussian Mixture Model

Adriana-Laura López-Lobato[1]([✉])[iD], Martha-Lorena Avendaño-Garrido[2][iD],
Héctor-Gabriel Acosta-Mesa[1][iD], José-Luis Morales-Reyes[3][iD],
and Elia-Nora Aquino-Bolaños[3][iD]

[1] Instituto de Investigaciones en Inteligencia Artificial, Universidad Veracruzana,
Xalapa-Enríquez, Veracruz, Mexico
adrilau17@gmail.com
[2] Facultad de Matemáticas, Universidad Veracruzana, Xalapa-Enríquez, Veracruz,
Mexico
[3] Centro de Investigación y Desarrollo en Alimentos, Universidad Veracruzana,
Xalapa-Enríquez, Veracruz, Mexico

Abstract. The classification of bean landraces based on their local variety and coloration is of particular interest because each bean landrace has a wide variety of nutritional components that benefit health. The laboratory procedures evaluate the diversity of local bean varieties based on color comparison as the primary element. In this paper, the authors propose a procedure to identify the coloration of the seeds according to the information provided for the CIE L*a*b* space of each bean landrace, a representation of the estimated color as a 3D histogram. This method uses this histogram to fit a statistical model known as the Gaussian Mixture Model with the GI algorithm. This process captures the representative information for each bean landrace as a point in the CIE L*a*b* space that represents the corresponding landrace color and can be used to identify the landraces using the K-nn method obtaining good results in homogeneous landraces.

Keywords: Bean landraces analysis · Gaussian Mixture Models · Gini index · CIE L*a*b* space · Optimization

1 Introduction

The common bean (Phaseolus vulgaris L.) is one of the most widely cultivated legumes due to its status as an important source of nutrition [2]. In Mexico and Central America, there is a great diversity of bean landraces that have been domesticated for the crop [4,12], since they have a high added value that takes advantage of the flower, green bean, and seeds [2,11].

The bean landraces are adapted to different conditions and agricultural practices in each region, such as different soil conditions and altitudes [14]. The great

L. Martínez-Villaseñor et al. (Eds.): MICAI 2024 Workshops, LNAI 15465, pp. 112–124, 2025.
https://doi.org/10.1007/978-3-031-83882-8_11

genetic diversity is composed of varieties of different coloration [9]. Grain colour, used as a phenotypic marker or selection criterion among farmers, is associated with differences in grain composition [8]. Proanthocyanidins, flavonol glycosides, and anthocyanins are responsible for the color of bean seeds, so they can be an indicator of bean health properties [5].

Dark bean landraces (black, red, or a mixture of dark grains) present higher anthocyanin content and antioxidant activity than white or yellow bean landraces [1]. For this reason, it is important to classify bean landraces and have a reference of their potential benefits for the health of consumers.

The diversity of bean landraces is evaluated by laboratory methods, where color is the main element of comparison. A spectrophotometer for solids is used to compare similarities between landraces, which is laboratory equipment that allows spot color measurement [1]. But a seed is not representative of a bean landrace, since the colorimetric representation of a landrace requires a set of seeds. Spot measurements are impractical to obtain the color distribution [1,7].

Computer vision systems have been successfully applied to seed classification [24]. Related work on colorimetric features reports the use of color averages to represent an individual seed. In addition, it is complemented by the seed color patch [15,20]. Since the color average is not representative, it is complemented with morphological information about size and shape [25], sometimes with Fourier elliptic descriptors for shape description, and texture descriptors of each bean sample [16], with principal component analysis of gray level histograms of a bean color region [23], or using the dominant color of each bean to characterize each bean [10]. Previous related work reports the bean classification from a single sample, not a set of seeds. In [19], a classification of bean landraces is reported using the color distribution of a set of 20 seeds to characterize the color with joint probability distributions and the K-nn method. In general, artificial intelligence techniques for classifying common bean landraces by color require a tool capable of classifying landraces that share functional similarities.

This work proposes a new approach to improve the classification accuracy of bean landraces by addressing a technique to identify the colorations that each of these landraces presents using the Gaussian mixture model (GMM) and the Gini index algorithm (GI algorithm). The GMM is a parametric model used to approximate the behaviour of a complex multimodal dataset. This is done by using a linear combination of simpler Gaussian distributions. Each Gaussian represents a different "cluster" within the analysed dataset [3]. The GI algorithm is an iterative process that estimates the parameters of a GMM for a given dataset [17,18].

Therefore, this paper aims to identify bean landraces based on the colorimetric properties of a set of seeds represented by a joint probability distribution and validate the performance of the classification method for 40 bean landraces.

The proposed method takes as input a 3D histogram in CIE L*a*b* space showing the normalized color frequencies of a bean landrace, and after processing the histogram information, the method fits a GMM with the GI algorithm. One of the key advantages of the GI algorithm over the EM algorithm, which is typically used to estimate the GMM parameters, is its capacity to identify the number of Gaussian components present in a given data set. If the bean landrace exhibits a single coloration (homogeneous color landrace), the GI algorithm will identify a single Gaussian component. If the bean landrace exhibits seeds of disparate colors (heterogeneous color landrace), the GI algorithm identifies as many Gaussian components as the number of colors. Subsequently, the GI algorithm identifies color distributions in the CIE L*a*b* space, conceptualized as a mixture of Gaussians. The corresponding means of these distributions serve as indicators of the colorations observed in the analyzed landrace. Subsequently, the means of all the landraces are utilized to classify them by color by applying the K-nearest neighbor (K-nn) method.

It is important to note that this methodology presents an "open set recognition problem" [22], since the histograms used in the training set, which serve as the basis for identifying primary colors in bean seeds, may not encompass all potential color variations observed in real-world scenarios. In other words, the training set may not fully represent the full range of colourations present in bean seeds. It is also worth mentioning that the results obtained with this process cant be compared with other methods, since the proposed methodology considers specific bean landraces with homogeneous colors.

The remaining paper is structured into four sections. In Sect. 2, a broad description of the color characterization of common bean landraces is shown. This section also presents an overview of the main characteristics of the Gaussian mixture model and the GI algorithm. Section 3 outlines the proposed methodology. Section 4 is dedicated to the experiments and results obtained. Finally, Sect. 5 presents the conclusions and future work.

2   Materials and Methods

2.1   Color Characterization of Common Bean Landraces

The common bean landraces were collected from rural communities in Oaxaca, Mexico. Only 40 bean landraces were included in the study due to the difficulty of sampling them because of their importance in the diet of local farmers [6]. Each local variety collected is represented by a 60-gram sample.

The illumination environment and color image reproduction workflow described in [19] were used to perform image acquisition and color calibration of the image set. In a digital image, the joint probability mass function was obtained by counting the frequency of occurrence of the three values of a pixel position in an image. Then, the joint probability of three discrete variables P_{lab} is given by the Eq. (1), where r_{lab} is the total number of occurrences (frequency) of the (l, a, b) pixel, and N is the number of pixels.

$$P_{lab} = \frac{r_{lab}}{N}. \tag{1}$$

Under this characterization, the three channels of the CIE L*a*b* color space were used. Each landrace colorimetric representation uses the color distribution of the seed set. The occurrence of different shades of color can be captured as the joint probability distribution of the pixel values of each channel [21].

The number of color histogram bins was calculated by discretizing the CIE L*a*b* channels into 256 values (2^8) since the range of the a* and b* channels is within $[-128, 127]$, for the L* channel, its values were normalized and scaled to 256 values. Therefore, the 3-dimensional histograms are expressed in the interval [0, 255] and are represented as a 3-dimensional $256 \times 256 \times 256$ matrix.

2.2   Gaussian Mixture Model and GI Algorithm

The Gaussian mixture model (GMM) is a parametric multimodal model to solve density estimation problems in unsupervised learning [3]. GMM considers a parametric density function that is a linear combination of K multivariate normal distributions f_{θ_k}, represented as in Eq. (2), where the mean μ_k and the covariance matrix Σ_k of the k-th Gaussian component are denoted by $\theta_k = (\mu_k, \Sigma_k)$, for $k = 1, \ldots, K$. The parameters ϕ_k are the weights or mixing proportions that must comply with $0 \le \phi_k \le 1$ for $k = 1, 2, \ldots, K$, and $\sum_{k=1}^{K} \phi_k = 1$.

$$\sum_{k=1}^{K} \phi_k f_{\theta_k}(\cdot), \text{ with } f_{\theta_k}(x) = \frac{1}{2\pi\sqrt{\Sigma_k}} e^{-\frac{1}{2}(x-\mu_k)^T \Sigma_k^{-1}(x-\mu_k)}. \tag{2}$$

The GMM is a statistical method that enables the modelling of dataset behaviour by identifying the value of the parameter $\theta = (\phi_k, \theta_k)_{k=1}^{K}$. One method for estimating the values of the GMM parameters for a given data set is the GI algorithm, which employs the Gini index as a basis for calculation [17,18].

The Gini index is the optimal value obtained through the resolution of a variant of the optimal transportation problem, wherein two probability distributions, ν_1 and ν_2, are considered within a random variable Y, and a distance d in Y is employed as a cost function, that is

$$GI(\nu_1, \nu_2) = \min_{\pi \in \Pi(\nu_1, \nu_2)} \left\{ \int_{Y \times Y} d(y_1, y_2) d\pi \right\}, \tag{3}$$

where $\Pi(\nu_1, \nu_2)$ is the set of joint probability distributions in $Y \times Y$ with marginals ν_1 and ν_2 in the first and second factors, respectively [13].

The Gini index, denoted by $GI(\nu_1, \nu_2)$, is a distance between the probability distributions ν_1 and ν_2 [26]. The Gini Index algorithm (GI algorithm) identifies the optimal parameters for a GMM by minimizing the distance given by the Gini index between the empirical distribution of the analyzed data and the density function of the GMM [17,18]. With these concepts, the authors define the GI algorithm for a data set $P = \{p_1, p_2, \ldots, p_M\}$, with $p_m = \left(p_1^{(m)}, p_2^{(m)}, \ldots, p_N^{(m)} \right) \in \mathbb{R}^N$, with the following expressions:

$$\mu_{tr} = \frac{\sum_{m=1}^{M} p_r^{(m)} \exp\left(-\frac{\left(p_r^{(m)} - \mu_{tr}\right)^2}{2\sigma_{tr}^2}\right)}{\sum_{m=1}^{M} \exp\left(-\frac{\left(p_r^{(m)} - \mu_{tr}\right)^2}{2\sigma_{tr}^2}\right)}, \quad \text{for } 1 \leq t \leq K \text{ and } 1 \leq r \leq N, \quad (4)$$

$$\sigma_{tr} = \frac{\sum_{m=1}^{M} \left(p_r^{(m)} - \mu_{tr}\right)^2 \exp\left(-\frac{\left(p_r^{(m)} - \mu_{tr}\right)^2}{2\sigma_{tr}^2}\right)}{\sum_{m=1}^{M} \exp\left(-\frac{\left(p_r^{(m)} - \mu_{tr}\right)^2}{2\sigma_{tr}^2}\right)}, \quad \text{for } 1 \leq t \leq K \text{ and } 1 \leq r \leq N,$$

$$(5)$$

where each mean μ_k is a real vector $(\mu_{k1}, \mu_{k2}, \ldots \mu_{kN})^T$ and the covariance matrix is a positive definite diagonal matrix of dimension $N \times N$, $\Sigma_k = diag\left(\sigma_{k1}^2, \sigma_{k2}^2, \ldots, \sigma_{kN}^2\right)$. In each iteration, the conditional probabilities, $P\left(f_{\theta_k} \mid p_m\right)$, for $k = 1, 2, 3, \ldots, K$, are employed to determine the probability that p_m originates from the parametric distribution k. Subsequently, the membership of each point is identified by considering the maximum conditional probability of that point and each Gaussian component. The mixture proportions ϕ_k are obtained by quantifying the elements in each class and normalizing them.

The favorable outcomes yielded by the experiments conducted in [17] and [18] with real and synthetic data under diverse circumstances of quantity, class intersection, dimensionality, and non-normal conditions substantiate the efficacy of the GI algorithm as a robust methodology for fitting a GMM.

3 Proposed Methodology

This section describes a methodology that employs the GI algorithm to compress the information of each bean landrace 3D histogram in the CIE L*a*b* space into a 3D point in the same space. Then, the landrace classification is performed using the K nearest neighbors (K-nn) method.

The following steps describe how to perform this information analysis.

1. Data cleansing.
2. Transfer of data to space $[0, 100] \times [0, 100] \times [0, 100]$.
3. Gaussian fitting with the GI algorithm.
4. Return mean values to the CIE L*a*b* space.

This procedure provides representative information about the analyzed bean landrace, given as the corresponding means in the CIE L*a*b* space. The landrace means in the CIE L*a*b* space obtained by this method are indicators of the colors of the corresponding landrace. By performing this process for each landrace, the information of all the landraces can be compared punctually by using a learning method, such as the K-nn method, and make the desired classifications in terms of seed color of bean landraces.

4   Experiments and Results

This section presents the experiments conducted on bean landraces with single colored seeds. Two categories are distinguished:

- **Primary class:** To identify each bean landrace with a label by its *local variety*. In this work, 40 different local varieties were considered, whose labels are listed in the legend of the Fig. 3.
- **Super class:** To study how the primary classes group together by their *color* similarities. The classification in terms of seed color is divided into five classes: white, yellow, red, brown, and black. Some examples of these colors' landraces are shown in Fig. 1.

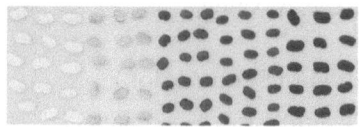

Fig. 1. Bean landraces with seed single color: white, yellow, red, brown, and black. (Color figure online)

This project involves 40 bean landraces. Each bean landrace has seeds of only one of the five colors mentioned above, i.e., each landrace has seeds of homogeneous color. In Table 1 are recorded the number of local varieties of each single color.

Table 1. Number of local varieties with seeds of a single color.

Number of local varieties	Color (Super class)
4	White
8	Yellow
9	Red
1	Brown
18	Black
40	Total

A sample of 10 CIE L*a*b* histograms for each of the bean landraces are obtained, for a total of 400 histograms.

In order to perform the classification, either by primary class or by superclass, the first step is to apply the methodology proposed in the previous section to obtain the corresponding values of the 400 means in the CIE L*a*b* space, and these values will correspond to the dataset. After obtaining a dataset with a sample size of 400, the following procedure will be repeated 1000 times:

1. The training and test sets for each bean landrace were randomly selected, with 70% of the data assigned to the training set and 30% assigned to the test set, i.e., select 7 out of 10 data points from each bean landrace for the training set, with the remaining 3 being assigned to the test set.

2. After performing this partition on each of the 40 bean landraces, a general training set of 280 items is obtained, 7 from each landrace, and a general test set of 120 items, 3 from each landrace.
3. The K-nn method is applied to classify the test set by using the general training set in the learning process, and the classification hits of the method are obtained. The knn() function from the class package of the free software R (www.r-project.org) is used for this purpose.
4. The achieved accuracy is recorded to analyze the results.

After conducting 1000 iterations of these experiments, the value of K used in the K-nn method, the mean and the variance of the classification hits, and the corresponding accuracy achieved are recorded. It is important to remember that the user must assign the K value.

4.1   Super Class Analysis (Seed Color)

This section presents the results of the seed color classification experiments conducted on bean landraces. An example of the training and test sets used for this classification is illustrated in Fig. 2.

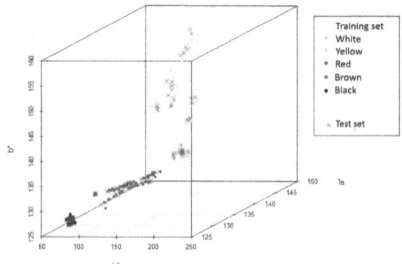

Fig. 2. Example of a training set and a test set for the super class analysis.

Table 2. Super class analysis results varying the K value in the K-nn method.

K value	Mean	Variance	Accuracy
1	120	0	100
2	120	0	100
3	120	0	100
4	120	0	100
5	120	0	100
6	120	0	100
7	120	0	100
8	120	0	100
9	120	0	100
10	120	0	100

The results presented in Table 2 were obtained after running the 1000 iterations of the process described in the previous section, using the values 1 to 10 for the K parameter in the K-nn method.

An accuracy of 100% is obtained with the proposed method for any K value in the K-nn method and for any training and test sets. In Fig. 2, it is observed that with the proposed methodology, the means in the CIE L*a*b* space are well differentiated with respect to the seed color, which leads to the high classification level achieved in this case.

4.2 Primary Class Analysis (Local Variety)

In these experiments, the classification of the 40 local varieties is considered. Figure 3 illustrates an example of the training and test sets used for this classification scheme. The results presented in Table 3 were obtained after executing the aforementioned process of 1000 iterations, using the values 1 to 10 for the K parameter in the K-nn method.

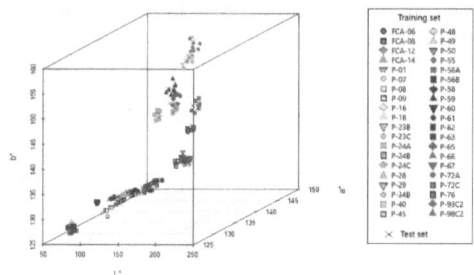

Fig. 3. Example of a training set and a test set for the analysis of the primary class.

Table 3. Primary class analysis results varying the K value in the K-nn method.

K value	Mean	Variance	Accuracy
1	77.09	14.94	64.25
2	75.74	14.70	63.12
3	78.36	15.09	65.30
4	78.03	13.44	65.02
5	78.05	13.24	65.04
6	77.57	13.04	64.64
7	77.35	15.87	64.46
8	76.51	16.08	63.76
9	75.41	15.45	62.84
10	74.61	16.75	62.17

Table 3 shows that with the proposed method, the percentages of success are between 62% and 65%, varying the parameter K value. It is important to remember that this percentage is obtained for the corresponding value of K after 1000 iterations. After analyzing each of the 10000 executions individually, i.e., the 1000 executions for each K value between 1 and 10, it is observed that the minimum value of hits is 64 and the maximum is 96 when 120 elements in the test set are observed, so the accuracy of the results varies between 53.33% and 80% in the 10000 iterations.

In the following section, a pre-classification by seed color is used to examine the classification by local variety, as this methodology always obtains satisfactory results in the classification by color. This pre-processing aims to determine the reasons for the low accuracy percentages observed in some seed colors.

4.3   Local Variety Analysis with a Pre-classification by Seed Color

In this section, the classification analysis by local variety (primary class) employs the pre-classification by seed color (super class) that was conducted previously. Consequently, the elements in the training and test sets can be identified by seed color, as illustrated in Table 4. Figure 4 depicts the training and test sets obtained for two seed colors: yellow and black. The following analysis will focus on the results obtained for these seed colors.

Table 4. Number of elements in the training and test sets by seed colors.

Local varieties	Seed color	Training set	Test set
4	White	28	12
8	Yellow	56	24
9	Red	63	27
1	Brown	7	3
18	Black	126	54

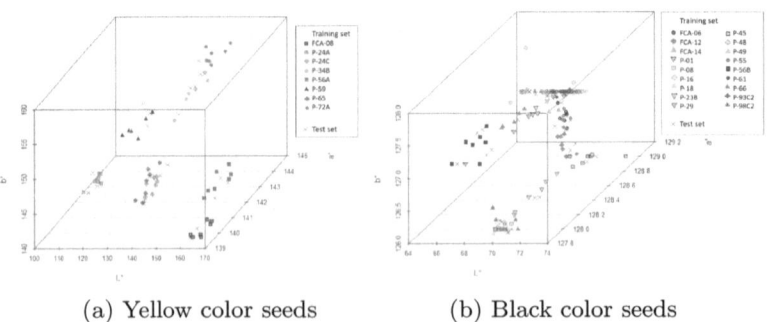

(a) Yellow color seeds   (b) Black color seeds

Fig. 4. Examples of training and test sets for yellow and black color seeds. (Color figure online)

Yellow Color Seeds. Figure 4(a) displays the eight local varieties of yellow-colored seeds. The K-nn method yielded the best result with $K = 3$, as shown in Table 5, after executing the 1000 experiments on the training and test elements of this seed color. By varying the value of K, an accuracy range between 79.17% and 100% is achieved, with an average accuracy of 93.52%, as between 19 and 24 correct answers out of 24 elements are observed in the test set.

Table 5. The best classification result by local variety using the K-nn method for the yellow-colored seeds.

K value	Mean	Variance	Accuracy
3	22.76	0.76	94.83

Black Color Seeds. As illustrated in Fig. 4(b), the black-colored seed has 18 local varieties, so there are 180 points of means, which are divided into training and test sets with 162 and 54 elements, respectively. A total of 10000 results were obtained from the 1000 iterations of the experiments using the K-nn method, with the K values varying from 1 to 10. The best result was obtained with $K = 1$, see Table 6.

Table 6. The best classification result by local variety using the K-nn method for the black-colored seeds.

K value	Mean	Variance	Accuracy
1	27.25	8.47	50.46

The number of correct answers varies between 17 and 37 classification hits, i.e. 31.5% and 68.5%. In this case, an average of 26.39 hits is obtained, i.e. an accuracy of 48.88%. The results for this color are not optimal. However, this type of result was expected due to the extensive number of local varieties examined, the small size of the test set, and the location of the means in the CIE L*a*b* space, as illustrated in Fig. 4(b).

A more detailed analysis by axis was performed on the black color seeds and it was determined that the axis that provides the most information about this color is L*. Figure 5 depicts the frequency histograms of the L*, a*, and b* axes, considering the means of this color without distinguishing between training and test sets. The information on the a* axis is concentrated within the values of 126 and 127. Graphically, the "lines" on these axes can be seen; see Fig. 4(b).

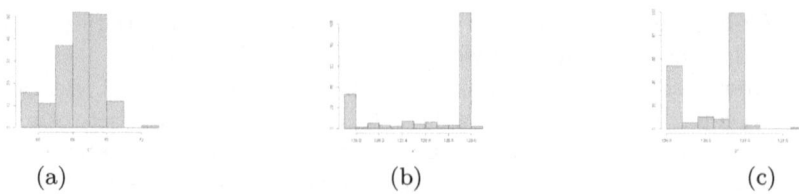

| (a) | (b) | (c) |

Fig. 5. Histograms of the L*, a* and b* axes of the values of the black color seeds means.

However, if the distinction between the local varieties is made using only the L* axis, which is representative of the information for this color, the information shown in Fig. 6 is obtained. Accordingly, the proposed methodology is inadequate for distinguishing between the various local varieties of black seeds, as their means in the CIE L*a*b* space are concentrated in a narrow area, leading to confusion in the classification. It is also noteworthy that experts have reported difficulties in differentiating between local varieties of black-colored seeds due to the high degree of similarity among bean landraces.

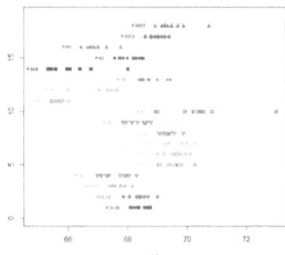

Fig. 6. L* axis intervals of the means in black color seeds.

5   Conclusions and Future Work

With the proposed methodology, favorable results are obtained in the classification of bean landraces with seeds of homogeneous color. The use of the GI algorithm with the information of the 3D histograms in the CIE L*a*b* space of the bean landraces shows well differentiated groups of means with respect to the seed color. Considering the classification by local varieties, it is evident that a pre-classification based on seed color yields superior outcomes for this task, as evidenced by the results obtained in this case. The results obtained from these experiments suggest that features other than seed color are needed to classify black seeds.

As future work, the proposed methodology will be used to classify bean landraces where the seeds do not have the same color and those where the seeds are mottled.

Acknowledgments. The first author is funded by the National Council of Humanities, Sciences and Technologies (CONAHCyT), through a postdoctoral fellowship at the Artificial Intelligence Research Institute of the University of Veracruz.

Disclosure of Interests. The authors have no competing interests to declare that are relevant to the content of this article.

References

1. Aquino-Bolaños, E.N., García-Díaz, Y.D., Chavez-Servia, J.L., Carrillo-Rodríguez, J.C., Vera-Guzmán, A.M., Heredia-García, E.: Anthocyanins, polyphenols, flavonoids and antioxidant activity in common bean (Phaseolus vulgaris L.) landraces. Emirates J. Food Agric. 581–588 (2016)
2. Aquino-Bolaños, E.N., et al.: Physicochemical characterization and functional potential of Phaseolus vulgaris L. and Phaseolus coccineus L. landrace green beans. Agronomy **11**(4), 803 (2021)
3. Bishop, C.M.: Pattern Recognition and Machine Learning. Springer (2006)
4. Bitocchi, E., et al.: Mesoamerican origin of the common bean (Phaseolus vulgaris L.) is revealed by sequence data. Proc. Natl. Acad. Sci. **109**(14), E788–E796 (2012)
5. Campos-Vega, R., Bassinello, P.Z., Santiago, R.A.C., Oomah, B.D.: Dry beans: processing and nutritional effects. In: Therapeutic, Probiotic, and Unconventional Foods, pp. 367–386 (2018)
6. Capistrán-Carabarin, A., Aquino-Bolaños, E.N., García-Díaz, Y.D., Chávez-Servia, J.L., Vera-Guzmán, A.M., Carrillo-Rodríguez, J.C.: Complementarity in phenolic compounds and the antioxidant activities of Phaseolus coccineus L. and P. vulgaris L. landraces. Foods **8**(8), 295 (2019)
7. Chávez-Mendoza, C., Hernández-Figueroa, K.I., Sánchez, E.: Antioxidant capacity and phytonutrient content in the seed coat and cotyledon of common beans (Phaseolus vulgaris L.) from various regions in Mexico. Antioxidants **8**(1), 5 (2018)
8. Chávez-Servia, J.L., et al.: Traditional family production and nutritional-nutraceutical value of common beans (Phaseolus vulgaris l.) in Southeast Mexico. In: Phaseolus Vulgaris: Cultivars, Production and Uses (2018)
9. Chávez-Servia, J.L., et al.: Diversity of common bean (Phaseolus vulgaris L.) landraces and the nutritional value of their grains. In: Grain Legumes, pp. 1–33. InTech, Rijeka (2016)
10. De Araújo, S.A., Pessota, J.H., Kim, H.Y.: Beans quality inspection using correlation-based granulometry. Eng. Appl. Artif. Intell. **40**, 84–94 (2015)
11. Espinosa-Pérez, E.N., Ramírez-Vallejo, P., Crosby-Galván, M.M., Estrada-Gómez, J.A., Lucas-Florentino, B., Chávez-Servia, J.L.: Clasificación de poblaciones nativas de frijol común del centro-sur de México por morfología de semilla. Rev. Fitotec. Mex. **38**(1), 29–38 (2015)
12. Espinoza-García, N., et al.: Contenido de minerales en semilla de poblaciones nativas de frijol común (Phaseolus vulgaris L.). Revista Fitotecnia Mexicana **39**(3), 215–223 (2016)

13. Gini, C.: Sulla misura della concentrazione e della variabilità dei caratteri. Atti del Reale Istituto veneto di Scienze, Lettere ed Arti **73**, 1203–1248 (1914)
14. Hernández-Delgado, S., et al.: Advances in genetic diversity analysis of Phaseolus in Mexico. In: Molecular Approaches to Genetic Diversity, vol. 1, pp. 47–73 (2015)
15. Kılıç, K., Boyacı, I.H., Köksel, H., Küsmenoğlu, İ: A classification system for beans using computer vision system and artificial neural networks. J. Food Eng. **78**(3), 897–904 (2007)
16. Koklu, M., Ozkan, I.A.: Multiclass classification of dry beans using computer vision and machine learning techniques. Comput. Electron. Agric. **174**, 105507 (2020)
17. López-Lobato, A.L., Avendaño-Garrido, M.L.: Using the gini index for a Gaussian mixture model. In: Martínez-Villaseñor, L., Herrera-Alcántara, O., Ponce, H., Castro-Espinoza, F.A. (eds.) MICAI 2020, Part II. LNCS (LNAI), vol. 12469, pp. 403–418. Springer, Cham (2020). https://doi.org/10.1007/978-3-030-60887-3_35
18. López-Lobato, A.L., Avendaño-Garrido, M.L.: Fitting a Gaussian mixture model through the gini index. Int. J. Appl. Math. Comput. Sci. **31**(3) (2021)
19. Morales Reyes, J.L., Acosta Mesa, H.G., Aquino Bolaños, E.N., Herrera Meza, S., Cruz Ramírez, N., Chávez Servia, J.L.: Classification of bean (Phaseolus vulgaris L.) landraces with heterogeneous seed color using a probabilistic representation. In: 2021 IEEE International Autumn Meeting on Power, Electronics and Computing (ROPEC), vol. 5, pp. 1–7. IEEE (2021)
20. Nasirahmadi, A., Behroozi-Khazaei, N., et al.: Identification of bean varieties according to color features using artificial neural network. Span. J. Agric. Res. **11**(3), 670–677 (2013)
21. Pishro-Nik, H.: Introduction to probability, statistics, and random processes (2016)
22. Scheirer, W.J., de Rezende Rocha, A., Sapkota, A., Boult, T.E.: Toward open set recognition. IEEE Trans. Pattern Anal. Mach. Intell. **35**(7), 1757–1772 (2012)
23. Tormena, C.D., Campos, R.C.S., Marcheafave, G.G., Bruns, R.E., Scarminio, I.S., Pauli, E.D.: Authentication of carioca common bean cultivars (Phaseolus vulgaris L.) using digital image processing and chemometric tools. Food Chem. **364**, 130349 (2021)
24. Velesaca, H.O., Suárez, P.L., Mira, R., Sappa, A.D.: Computer vision based food grain classification: a comprehensive survey. Comput. Electron. Agric. **187**, 106287 (2021)
25. Venora, G., Grillo, O., Ravalli, C., Cremonini, R.: Identification of Italian landraces of bean (Phaseolus vulgaris L.) using an image analysis system. Scientia Horticulturae **121**(4), 410–418 (2009)
26. Villani, C.: Optimal Transport: Old and New, vol. 338. Springer (2008)

Efficient Neural Architecture Search: Computational Cost Reduction Mechanisms in DeepGA

Jesús-Arnulfo Barradas-Palmeros$^{(\boxtimes)}$ ⓘ, Carlos-Alberto López-Herrera ⓘ,
Héctor-Gabriel Acosta-Mesa ⓘ, and Efrén Mezura-Montes ⓘ

Artificial Intelligence Research Institute, University of Veracruz, Xalapa, Mexico
{zS23000652,zS23000650}@estudiantes.uv.mx, {heacosta,emezura}@uv.mx

Abstract. Neural Architecture Search (NAS) aims to automate the design process of Deep Neural Networks (DNN) without requiring profound domain knowledge. The Deep Genetic Algorithm (DeepGA) was proposed to find the architectures of Convolutional Neural Networks (CNNs) for image processing, and its applications have covered a variety of data domains. Nonetheless, one of the main impediments of NAS is its computational cost, which is produced by evaluating candidate architectures. This work proposes using two cost-reduction mechanisms applied to DeepGA: using memory to avoid repeated evaluations and reducing the number of epochs for training in a low-fidelity estimation scheme for evaluations. The results indicate that the execution time of the algorithm was extensively reduced without diminishing the accuracy performance of the resulting architecture. The previous derives in a more efficient NAS procedure.

Keywords: Neural Architecture Search · Cost reduction · Convolutional Neural Networks

1 Introduction

Convolutional Neural Networks (CNNs), as a Deep Learning (DL) model, have achieved noteworthy performance in computer vision tasks, such as image classification [11]. CNNs comprise convolutional, pooling, and fully connected layers where several hyperparameters must be defined [23]. Traditionally, the design of CNNs and other types of Deep Neural Networks (DNN) requires domain knowledge and DL expertise, making their application difficult on new problems. Neural Architecture Search (NAS) counters the previous problem by developing an automatic process to construct competitive problem-tailored models [14].

As stated in [7], NAS has three main components: the search space, the search strategy, and the performance estimation method. The search space is defined by the network representation called neural encoding [18]. The search strategy is how the search space is explored to find the appropriate network architecture.

© The Author(s), under exclusive license to Springer Nature Switzerland AG 2025
L. Martínez-Villaseñor et al. (Eds.): MICAI 2024 Workshops, LNAI 15465, pp. 125–134, 2025.
https://doi.org/10.1007/978-3-031-83882-8_12

Evolutionary Computation (EC) techniques have been widely used in this case [2]. Finally, the performance estimation consists of the method used to evaluate an architecture. One of the main limitations of NAS is the computational time required in the search process using specialized hardware, including Graphics Processing Units (GPUs). Typically, an architecture is trained and evaluated in a resource-intensive process. Nevertheless, acceleration techniques have been proposed, such as surrogate models, training-free NAS, and low-fidelity evaluations [6,21].

A surrogate model incorporates a lower-cost model into the search process to replace some expensive evaluations. The surrogate model is trained with pairs of network encoding and its fitness value. Then, it is used to predict the performance of candidate architectures [8]. In training-free NAS, a score function, also known as zero-shot evaluation, is used to estimate the performance of a candidate network as an alternative to the classic training and evaluation process [21]. Lastly, a low-fidelity evaluation attempts to accelerate an evaluation by reducing the number of epochs used for training or lightening the data used for training. The latter includes using low-resolution images, smaller proxy datasets, or subsets of the training set [6]. The number of epochs used to train a candidate architecture directly impacts resource consumption. In [15], the search process is conducted using a reduced number of epochs for training under the assumption that identifying the tendency of the performance is enough to estimate which architectures are better than others. At the end of the search process, the best neural architecture found is trained for an additional number of epochs.

When EC algorithms are used for NAS, memory can be used to avoid conducting repeated evaluations in the search process. Consequently, resource demand is reduced. The first time an individual appears and is evaluated, its fitness value is stored in memory. If the evolutionary process creates an individual equal to one previously evaluated, the fitness value is returned from memory instead of reevaluating it [9]. This simple mechanism's capabilities for computational cost reduction have been evaluated in [1]. In NAS, examples where memory has been used are presented in [16] with a Genetic Algorithm (GA) and in [5] with Genetic Programming (GP).

The Deep Genetic Algorithm (DeepGA) was proposed in [17] to automatically design CNNs in single- and multi-objective versions, considering network performance and complexity. DeepGA uses a hybrid encoding to represent the networks. Dense and fully connected blocks represent the convolutional and fully connected layers, respectively. Skip connections are represented with a binary string. The algorithm was proven successful at generating well-performing architectures with fewer parameters when tested with a dataset of chest X-ray images. Later, DeepGA has been applied in the designing of CNNs architectures for breast cancer diagnosis [10], vehicle make and model recognition [20], estimation of Anthocyanins in bean landraces [12,13], and steering angle estimation [19].

Dealing with the computational cost of the NAS approaches is of uttermost importance for future applications, especially if access to specialized hardware is limited. This work applies cost-reduction strategies to DeepGA, a NAS procedure tested in various domains. First, memory is included to avoid repeated evaluations. After that, the number of training epochs is reduced following a low-fidelity evaluation estimation scheme. Reducing the algorithm's computational time is expected, and the algorithm's performance in terms of classification accuracy and number of parameters is analyzed. Alternative techniques, such as surrogate models and training-free NAS approaches, involve drastically changing the overall sequence of the search and evaluation processes, respectively. Contrarily, the selected mechanisms for cost-reduction require simple algorithm modifications and serve as an initial approach to make DeepGA more efficient.

The rest of the paper is organized into four sections. Section 2 introduces DeepGA. The experimentation details and the proposed configurations for computational cost reduction are presented in Sect. 3. In Sect. 4, the results are shown. Finally, Sect. 5 includes the conclusions and states areas identified for future work.

2   The Deep Genetic Algorithm

DeepGA was introduced in [17] as a procedure for designing CNNs automatically. A GA is used to evolve candidate architectures following a classic evolutionary scheme: an initial population is created, parent selection is conducted, variation operators (crossing and mutation) are applied, and environmental selection is performed. The last three steps are repeated until a maximum number T of iterations (generations) are completed. More details on the algorithm are presented in the rest of this section.

2.1   Encoding and Initial Population

As seen in [18], an encoding represents a neural network. It can be classified as direct, indirect, and hybrid according to the way the characteristics of the network are represented, as well as fixed or variable length. DeepGA uses a hybrid encoding composed of convolutional and fully connected blocks (direct) and a binary string representing the skip connections (indirect). The encoding has a variable length since it can grow to represent more architecture layers without a predefined architecture size.

In the first level encoding, the convolutional blocks include the number of filters and the filter size as evolvable hyperparameters. In addition, the blocks can be used with no pooling, max pooling, or average pooling. The pooling type is encoded as part of the convolutional block. If pooling is enabled (max or average), the kernel size is also included as an evolvable hyperparameter. A stride of 1 is always used in the convolutional layers with zero padding. ReLU is used as an activation function, and batch normalization is applied. The stride is defined as 2 for pooling. The fully connected blocks contain the number of neurons of the layer as an evolvable hyperparameter. ReLU is also used.

The second label encoding, composed of the binary string, represents the skip connections starting in the third convolutional layer. A value of one indicates that a skip connection exists, and zero represents the opposite. For example, the string 100101 describes the skip connections in a network with five convolutional layers. The third layer receives a skip connection from the first layer, the fourth layer does not receive skip connections, and the fifth layer receives connections from the first and third layers. If the skip connections need dimensionality adjustments, zero padding or max pooling is applied depending on the spatial resolution adjustment required. The decoding of an individual is shown in Fig. 1.

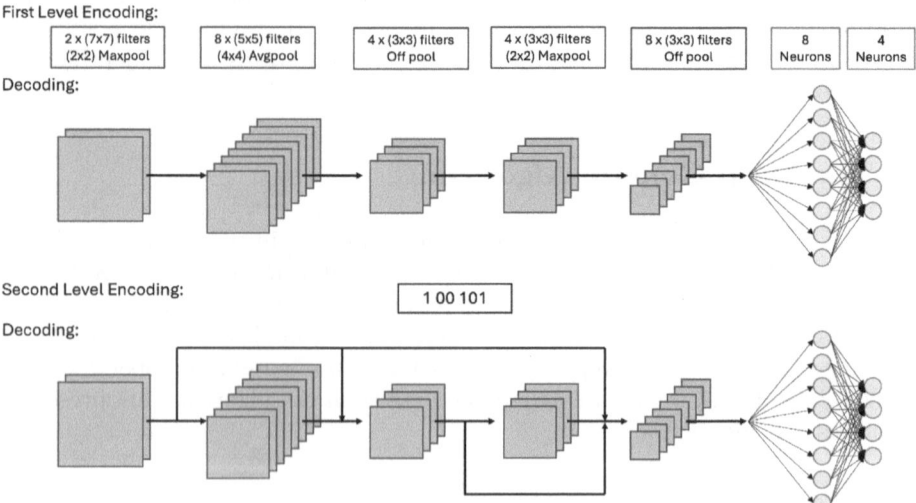

Fig. 1. An example of an individual encoding and decoding process. The first level encoding uses blocks to represent the convolutional and fully connected layers. In the second level, skip connections are represented with a binary string.

DeepGA uses a population of size N. At the beginning of the procedure, the population is initialized with a random selection of the number of convolutional (from 2 to 5) and fully connected layers (from 1 to 4). Then, for each convolutional layer, random choices are applied to select the number of filters, filter size, type of pooling operation, and kernel size for pooling. Afterward, the number of neurons is randomly selected for each fully connected layer. For the second level encoding, a binary string is generated randomly.

2.2   Selection Mechanisms

A stochastic tournament is used to select individuals from the population as parents. The tournament size t_s is a user-defined parameter that changes the selection pressure. The tournament takes place by choosing t_s individuals from the population at random. The individual with the best fitness value is selected with a probability of 80%. Otherwise, the parent is selected randomly from the individuals that form the tournament. The crossover and mutation operators are applied in each generation until $N/2$ individuals are generated. The offspring is then added to the population. Elitism is applied, maintaining the best N individuals in the population for the next generation.

2.3   Crossover and Mutation Operators

The crossover operator, with a probability defined by the user as the crossover rate (CR), combines two parents to create two descendants. Crossover in DeepGA is applied on each level of the encoding. First, the parent with less convolutional blocks is identified as P_1 and the other as P_2. m is defined as the floor function of the number of convolutional blocks in P_1 divided by two. The parents' last m convolutional blocks are exchanged. For the second level encoding, c is defined as the floor function of dividing the length of P_1 binary string by two. The last c bits of P_1 and P_2 are exchanged. Then, the parent with less fully connected blocks is now identified as P_1, and n is calculated as the floor function of dividing by two the number of fully connected blocks in P_1. Afterward, the parents' last n fully connected blocks are exchanged.

Once a descendant is generated, it goes through a mutation process. The mutation is applied with a user-defined probability called the mutation rate (MR). Two forms of mutation are considered in DeepGA. r_1 and r_2 are randomly generated numbers from a uniform distribution in the range of $(0, 1)$. If $r_1 \leq 0.5$, the first mutation is applied. In this case, a block is restarted, and the value of a bit chosen randomly in the binary string is flipped. If $r_2 > 0.5$, a fully connected block is restarted, choosing a new number of neurons. Otherwise, a convolutional block is restarted, randomly choosing the number of filters, the filter size, the type of pooling, and the size of the pooling kernel.

When $r_1 > 0.5$, the second form of mutation is used. If $r_2 > 0.5$, a fully connected block is added in a random position; otherwise, the added block is convolutional. The addition of a fully connected block does not affect the second-level encoding. On the other hand, when a new convolutional block is attached to the individual, new bits need to be included in the second-level encoding. The position where the new block was inserted determines where the additional bits must be appended. The bit values are assigned randomly.

2.4   Fitness Function

DeepGA considers the CNNs architecture's performance and complexity when performing an evaluation. The classification accuracy metric is considered to measure the network performance. On the other hand, the number of network parameters is used for the complexity. The algorithm can use two optimization approaches: single- and multi-objective. In the single objective framework, both objectives are considered in a single weighted fitness function as presented in Eq. 1.

$$f(cnn) = (1 - w) * accuracy(cnn) + w * \frac{MP - NP}{MP} \tag{1}$$

where cnn represents the architecture encoded by an individual. w is used to weigh the considered objectives. A higher value of w will increase the importance of the architecture's complexity in fitness calculation. MP is a parameter defined as a candidate architecture's upper limit in the number of parameters. MP is set as 2×10^6. NP is the number of parameters of cnn. In the DeepGA's multi-objective version, the objectives are considered as two independent functions, and the selection mechanism to reduce the population after offspring calculation is changed. Pareto dominance is considered, and the population is sorted according to the resulting fronts from applying the non-dominated sort algorithm from [4].

3   Experimentation

Two changes are applied in the single-objective version of DeepGA to reduce its computational cost. First, memory is used to avoid conducting repeated evaluations. In this case, when the procedure requires evaluating an individual, it first searches in memory for an existing entry associated with the individual. If the individual has not been previously evaluated, the procedure performs the evaluation and stores the individual and its fitness value in memory. Avoiding repeated evaluations is expected to directly impact the method's execution time since the training process of a network is performed on fewer occasions.

The second change is to reduce the computational time required to conduct an evaluation by reducing the number of epochs for training. This proposal is based on the assumption that knowing the performance tendency is enough to conduct the search process. Initially, DeepGA used 10 epochs to train an individual. This parameter has been maintained or augmented in the works where DeepGA is applied: 10 epochs in [10], 20 epochs in [20], and 30 epochs in [12,13,19]. This work reduces the number of epochs to train an individual to 5, and the impact on the algorithm's performance is evaluated.

Four algorithm configurations are tested to study the effects of the cost-reduction mechanisms applied. First, each mechanism is applied separately to evaluate its particular effect. Then, the mechanisms are combined in a proposal. The proposed configurations are the following:

1. DeepGA: the original DeepGA procedure using 10 epochs.
2. Configuration 1: DeepGA with memory to avoid repeated evaluations. 10 epochs are used.
3. Configuration 2: DeepGA with the reduced number of epochs. 5 epochs are used.
4. Configuration 3: DeepGA with memory to avoid repeated evaluations and using 5 epochs for training.

The best architecture found by the proposals at the end of the search process is trained for 50 epochs. Five runs of each configuration are performed using an Nvidia T4 GPU in the Google Colab pro+ virtual environments. The fashion MNIST dataset [22] is selected for experimentation. The data contains 28×28 gray-scale images belonging to ten different clothing items. The training set contains 60,000 images, whereas the test set contains 10,000 images. Given the small dimensions of the data, some parameters of DeepGA are adjusted. The parameter configuration is presented in Table 1. Adam optimizer is used for training with a learning rate of 1×10^{-4}.

Table 1. Parameter configuration for DeepGA.

Network hyperparameters		GA parameters	
Conv. filter size	{2, 3, 4}	N	20
No. of conv filters	{2, 3, 4, 16, 32}	T	40
Pooling type	{Max, Avg, off}	CR	0.7
Pooling kernel size	{2, 3}	MR	0.5
No. of neurons FC	{4, 8, 16, 32, 64, 128}	t_s	5

4   Results

First, the performance obtained with the proposed configurations of DeepGA is analyzed. Table 2 presents the mean accuracy results obtained after training the best individual for 50 epochs and the mean complexity. Statistical tests are conducted with the accuracy values obtained with the different configurations. Initially, the results were tested for normality using the Shapiro-Wilk test. In the configuration 2, normality is discarded. Then, the Friedman non-parametric test was applied. With a p-value of 0.35, the test did not detect significant differences among the results of the tested configurations. An interesting tendency is seen in the number of parameters of the resulting architectures. The configurations that used fewer epochs for training preferred more complex networks.

The number of evaluations and the execution time were monitored to test the effectiveness of the cost-reduction mechanisms. The memory mechanism is expected to reduce the total number of evaluations the method performs. Selecting a smaller number of epochs for training directly affects the computational time since the evaluations are performed faster. Table 3 presents the mean number of evaluations and execution time of tested configurations. The original DeepGA procedure is the basis for comparison, performing 420 evaluations and taking 22.35 h to complete. As seen in the results from configurations 2 and 3, the cost-reduction mechanisms were able to reduce more than 40% of the computational time when applied independently. When applied in combination, as in configuration 3, the time reduction was increased to 69.09%.

Table 2. Performance and complexity of the resulting architectures found by the proposed configurations of DeepGA. The mean values are presented, the best result is marked in bold, and a ranking among the tested variants is included.

Method	No. of Params			Accuracy		
	Mean	Std dev	Rank	Mean	Std dev	Rank
DeepGA	42034.0	9054.5	(2)	0.9133	0.0064	(4)
Conf1	**36540.4**	10928.6	(1)	**0.9186**	0.0050	(1)
Conf2	61429.2	19468.3	(4)	0.9138	0.0020	(3)
Conf3	55039.6	17593.3	(3)	0.9174	0.0065	(2)

Table 3. Mean number of evaluations and computational time reported by the different configurations tested. The reduction considering the original DeepGA method is included. The best result is marked in bold.

Method	No. of Evals	Variation	Time (hours)	variation
DeepGA	420.0	–	22.35	–
Conf1	239.2	43.05%	12.95	42.05%
Conf2	420.0	0.00%	11.99	46.33%
Conf3	232.0	**44.76%**	6.91	**69.09%**

The use of the memory mechanism and the reported number of evaluations performed by the method indicate that more than 40% of the architectures found with the algorithm's variation operators were previously explored. The findings can be used for parameter tuning in future DeepGA applications.

5   Conclusions and Future Work

This paper applied two cost-reduction mechanisms to DeepGA. The first one corresponds to the use of memory to avoid repeated evaluations. The second one involved applying a low-fidelity estimation process, reducing the number of epochs to train an individual during an evaluation. The results show that, on average, 44.76% of the evaluations were avoided, and the computational time had a 69.09% reduction. The algorithm's performance was maintained following a more efficient process, yet the number of architecture parameters was increased. Nonetheless, these findings are limited to one dataset and five procedure runs. Additional experimentation is needed to evaluate if this tendency holds when applying DeepGA to other data.

By reducing the execution time of DeepGA, the associated energy consumption is directly diminished. Improving the algorithm's efficiency aspires to decrease its environmental impact, which is aligned with the Green Artificial Intelligence trend [3]. In future experimentation, other strategies for computational cost reduction can be used with DeepGA. Attractive alternatives include a training-free NAS process with a performance predictor for the network evaluation or incorporating a surrogate model. Modifications to the encoding used in DeepGA can also be explored to increase the performance. Finally, as presented in [6,23], considering explainability mechanisms is a future direction for NAS algorithms. This way, the resulting architecture can provide insights into its decision-making process.

Acknowledgments. The first (CVU 1142850) and second (CVU 1075919) authors acknowledge Mexico's National Council of Humanities, Science, and Technology (CONAHCYT) for the scholarships awarded for graduate studies at the University of Veracruz.

References

1. Barradas-Palmeros, J.A., Mezura-Montes, E., Rivera-López, R., Acosta-Mesa, H.G.: Computational cost reduction in wrapper approaches for feature selection: a case of study using permutational-based differential evolution. In: 2024 IEEE Congress on Evolutionary Computation (CEC), pp. 1–8 (2024). https://doi.org/10.1109/CEC60901.2024.10611859
2. Bi, Y., Xue, B., Mesejo, P., Cagnoni, S., Zhang, M.: A survey on evolutionary computation for computer vision and image analysis: past, present, and future trends. IEEE Trans. Evol. Comput. **27**(1), 5–25 (2023). https://doi.org/10.1109/TEVC.2022.3220747
3. Bolón-Canedo, V., Morán-Fernández, L., Cancela, B., Alonso-Betanzos, A.: A review of green artificial intelligence: towards a more sustainable future. Neurocomputing **599**, 128096 (2024)
4. Deb, K., Pratap, A., Agarwal, S., Meyarivan, T.: A fast and elitist multiobjective genetic algorithm: NSGA-II. IEEE Trans. Evol. Comput. **6**(2), 182–197 (2002). https://doi.org/10.1109/4235.996017
5. Fuentes-Tomás, J.A., Mezura-Montes, E., Acosta-Mesa, H.G., Márquez-Grajales, A.: Tree-based codification in neural architecture search for medical image segmentation. IEEE Trans. Evol. Comput. **28**(3), 597–607 (2024). https://doi.org/10.1109/TEVC.2024.3353182
6. Li, N., et al.: Automatic design of machine learning via evolutionary computation: a survey. Appl. Soft Comput. **143**, 110412 (2023)
7. Li, Y., Liu, J.: A survey: evolutionary deep learning. Soft. Comput. **27**(14), 9401–9423 (2023)
8. Liang, J., Lou, Y., Yu, M., Bi, Y., Yu, K.: A survey of surrogate-assisted evolutionary algorithms for expensive optimization. J. Membr. Comput. (2024)
9. Liu, Y., Sun, Y., Xue, B., Zhang, M., Yen, G.G., Tan, K.C.: A survey on evolutionary neural architecture search. IEEE Trans. Neural Netw. Learn. Syst. **34**(2), 550–570 (2023). https://doi.org/10.1109/TNNLS.2021.3100554
10. Llaguno-Roque, J.L., Barrientos-Martínez, R.E., Acosta-Mesa, H.G., Romo-González, T., Mezura-Montes, E.: Neuroevolution of convolutional neural networks for breast cancer diagnosis using western blot strips. Math. Comput. Appl. **28**(3) (2023). https://doi.org/10.3390/mca28030072. https://www.mdpi.com/2297-8747/28/3/72

11. Mishra, V., Kane, L.: A survey of designing convolutional neural network using evolutionary algorithms. Artif. Intell. Rev. **56**(6), 5095–5132 (2023)
12. Morales-Reyes, J.L., Aquino-Bolaños, E.N., Acosta-Mesa, H.G., Márquez-Grajales, A.: Estimation of anthocyanins in homogeneous bean landraces using neuroevolution. In: Calvo, H., Martínez-Villaseñor, L., Ponce, H., Zatarain Cabada, R., Montes Rivera, M., Mezura-Montes, E. (eds.) Advances in Computational Intelligence. MICAI 2023 International Workshops, pp. 373–384. Springer, Cham (2024)
13. Morales-Reyes, J.L., Aquino-Bolaños, E.N., Acosta-Mesa, H.G., Márquez-Grajales, A.: Estimation of anthocyanins in heterogeneous and homogeneous bean landraces using probabilistic colorimetric representation with a neuroevolutionary approach. Math. Comput. Appl. **29**(4) (2024). https://doi.org/10.3390/mca29040068. https://www.mdpi.com/2297-8747/29/4/68
14. Poyser, M., Breckon, T.P.: Neural architecture search: a contemporary literature review for computer vision applications. Pattern Recogn. **147**, 110052 (2024). https://doi.org/10.1016/j.patcog.2023.110052. https://www.sciencedirect.com/science/article/pii/S0031320323007495
15. Sun, Y., Xue, B., Zhang, M., Yen, G.G.: Evolving deep convolutional neural networks for image classification. IEEE Trans. Evol. Comput. **24**(2), 394–407 (2020). https://doi.org/10.1109/TEVC.2019.2916183
16. Sun, Y., Xue, B., Zhang, M., Yen, G.G., Lv, J.: Automatically designing CNN architectures using the genetic algorithm for image classification. IEEE Trans. Cybern. **50**(9), 3840–3854 (2020). https://doi.org/10.1109/TCYB.2020.2983860
17. Vargas-Hákim, G.A., Mezura-Montes, E., Acosta-Mesa, H.G.: Hybrid encodings for neuroevolution of convolutional neural networks: a case study. In: Proceedings of the Genetic and Evolutionary Computation Conference Companion, GECCO '21, pp. 1762–1770. Association for Computing Machinery, New York (2021). https://doi.org/10.1145/3449726.3463133
18. Vargas-Hákim, G.A., Mezura-Montes, E., Acosta-Mesa, H.G.: A review on convolutional neural network encodings for neuroevolution. IEEE Trans. Evol. Comput. **26**(1), 12–27 (2022). https://doi.org/10.1109/TEVC.2021.3088631
19. Velazco-Muñoz, J.D., Acosta-Mesa, H.G., Mezura-Montes, E.: Reducing parameters byÂ neuroevolution inÂ cnn forÂ steering angle estimation. In: Mezura-Montes, E., Acosta-Mesa, H.G., Carrasco-Ochoa, J.A., Martínez-Trinidad, J.F., Olvera-López, J.A. (eds.) Pattern Recognition, pp. 377–386. Springer, Cham (2024)
20. Vázquez-Santiago, D.I., Acosta-Mesa, H.G., Mezura-Montes, E.: Vehicle make and model recognition as an open-set recognition problem and new class discovery. Math. Comput. Appl. **28**(4) (2023). https://doi.org/10.3390/mca28040080. https://www.mdpi.com/2297-8747/28/4/80
21. Wu, M.T., Tsai, C.W.: Training-free neural architecture search: a review. ICT Express **10**(1), 213–231 (2024)
22. Xiao, H., Rasul, K., Vollgraf, R.: Fashion-MNIST: a novel image dataset for benchmarking machine learning algorithms (2017). https://arxiv.org/abs/1708.07747
23. Zhan, Z.H., Li, J.Y., Zhang, J.: Evolutionary deep learning: a survey. Neurocomputing **483**, 42–58 (2022)

Prediction of Epileptic Seizure Using Neuroevolved Spiking Neural Network

Carlos-Alberto López-Herrera[✉][iD], Héctor-Gabriel Acosta-Mesa[iD],
Efrén Mezura-Montes[iD], and Jesús-Arnulfo Barradas-Palmeros[iD]

Artificial Intelligence Research Institute, University of Veracruz, Xalapa, Mexico
{zs23000650,zs23000652}@estudiantes.uv.mx, {heacosta,emezura}@uv.mx
https://www.uv.mx/iiia/

Abstract. Epilepsy is a worldwide common neurological condition where patients suffer from recurrent seizures, which are brief episodes of involuntary movement usually accompanied by loss of consciousness. To predict these events, researchers have analyzed the transition between two epileptic brain signals: interictal and pre-ictal. This work proposes to apply the NeuroEvolution of Augmenting Topologies (NEAT) as a neuroevolution approach to obtain optimal Spiking Neural Networks (SNNs) as a model to predict epileptic seizures. The CHB-MIT database, which collects brain epileptic signals obtained by the Electroencephalography (EEG) technique, was selected to test the model. The sensibility, specificity, and Seizure Prediction Horizon (SPH) metrics were used. The latter assesses the capacity to forecast an episode. The method showed competitive results, reaching mean values of 94.39% in sensibility, 81.65% in specificity, and 44.73 min in SPH, which are in the range of state-of-the-art proposals. Furthermore, our best and median results increased the SPH, achieving 50.34 and 48.10 min, respectively.

Keywords: Epileptic seizure · Spiking Neural Networks · NEAT

1   Introduction

According to the World Health Organization (WHO), epilepsy is one of the most common neurological disorders affecting around 50 million people worldwide, of which 80% live in low- and middle-income countries [1]. The social and economic impact is significant, as epilepsy can lead to stigma, decreased quality of life, and financial burdens due to medical costs and loss of productivity [2]. Epilepsy is defined as a chronic noninfectious disease of the brain characterized by recurrent seizures, which are brief episodes of involuntary movement usually accompanied by loss of consciousness [1,3]. Seizures can be focal, originating in a specific area of the brain, or generalized, affecting both hemispheres. Since these episodes result from excessive electrical discharges in a group of brain cells, methods that record the electrical neural activity are often used to study this disorder. Such is the case of the electroencephalography (EEG).

© The Author(s), under exclusive license to Springer Nature Switzerland AG 2025
L. Martínez-Villaseñor et al. (Eds.): MICAI 2024 Workshops, LNAI 15465, pp. 135–146, 2025.
https://doi.org/10.1007/978-3-031-83882-8_13

The EEG is an electrophysiological method that allows recording the brain's electrical activity in real-time. Some of its more proficient advantages are its high temporal sensibility, low cost, and portability [4]. For standardized measurements, EEG signals are typically recorded using the 10–20 system (Fig. 1), a widely adopted method that specifies the placement of all sensors (electrodes) based on anatomical landmarks on the scalp. This practice ensures consistent and reproducible monitoring of brain activity in distinct channels across different individuals and studies.

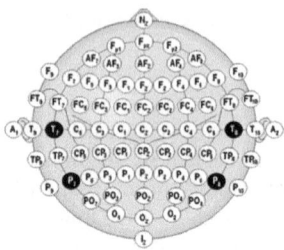

Fig. 1. The 10–20 system. Letters convey the position according to each brain lobe (F for frontal, P for parietal, T for temporal, O for occipital, and C for central). Odd and even numbers are used for left and right positions, respectively [4].

By the electrical activity, it is possible to characterize epileptic brain signals into four states (Fig. 2): interictal, pre-ictal, ictal, and post-ictal [3]. First, interictal activity refers to the phase between two epileptic episodes. Furthermore, the pre- and ictal phases signify the periods before and during the seizure. Finally, the post-ictal represents the period after the event. With this, the transition between interictal and pre-ictal marks the beginning of an epileptic seizure episode [5].

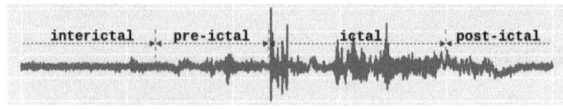

Fig. 2. Stages of characteristic epileptic brain signals [6].

Several Artificial Intelligence (AI) methods have been proposed to predict an epileptic seizure from EEG signals [5,7]. Nevertheless, those efforts have not focused on the temporal aspect of the signals. Spiking Neural Networks, the third generation of Artificial Neural Networks (ANNs), excel at recognizing temporal patterns emerging from time-series signals [8,9].

The SNN model is a closer approximation of the brain than traditional ANNs, producing discrete impulses (spikes) rather than continuous values. Therefore, SNNs mimic a biological neuron by emitting and distributing pulses at specific moments [10]. In that sense, SNNs are more capable of handling temporal information and variations [9]. SNNs use realistic biological models as nodes (computational units) to produce spikes. Given its simplicity and optimal operation, the Leaky Integrate-and-Fire (LIF) is one of the most used in current literature [11]. The LIF model uses spike trains as inputs to modify its inner membrane potential, spiking when a threshold has been reached (Fig. 3). Mathematically, the LIF model, proposed in 1967 by Stein [12], is shown in Eqs. 1 and 2.

$$\tau_m \frac{dV}{dt} = -(V - E_L) + I(t)R_m \tag{1}$$

$$\begin{aligned} V &= V_{\text{rest}} \quad \text{if} \quad V < V_{\text{rest}} \\ \text{if} \quad V &\geq V_{\text{th}}, \quad V \rightarrow V_{\text{reset}}, \quad \text{and} \quad t \rightarrow t + \Delta t_{\text{ref}} \end{aligned} \tag{2}$$

where V is the membrane potential, E_L is the resting potential, $I(t)$ is the input current, R_m is the membrane resistance, τ_m is the membrane time constant, V_{rest} is the resting potential, V_{th} is the threshold potential, V_{reset} is the reset potential, and Δt_{ref} is the refractory period [13].

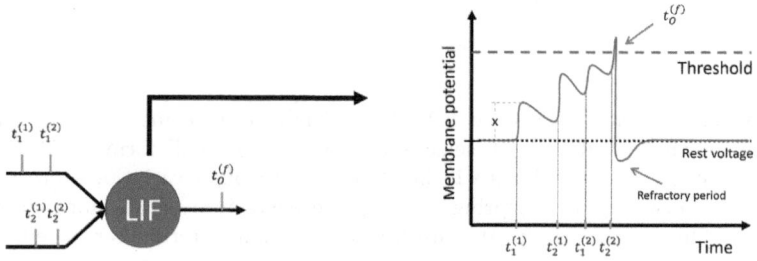

Fig. 3. An LIF neuron connected to two inputs t_1 and t_2, which receives spikes at specific times (1 and 2). Internally, the neuron modifies its membrane potential by a specific factor x whenever a spike is received, increasing or decreasing it if the connection is excitatory or inhibitory. When enough spikes are received to reach a threshold value, a neuron fires a spike t_O to all outgoing connections. Afterward, the neuron starts a refractory period, blocking all incoming inputs, to then return to its initial resting value.

Designing and training SNNs is a laborious and highly complex task. Unlike traditional ANNs, synaptic weights cannot be trained using standard optimization methods used in AI, such as gradient descent, due to the non-differentiability nature of the neuron models. While some techniques, such as surrogate gradient descent [14], have been proposed, they have not yet achieved competitive effectiveness due to limitations like non-exact gradient approximations and computational inefficiencies [15]. These challenges underscore the critical need for using automated optimization techniques in this field.

Neuroevolution (NE) uses Evolutionary Algorithms (EAs) (Fig. 4) as a search method to find optimal neural networks for a given task. Essential characteristics that may be optimized through NE include the architecture, synaptic weights, and neuron properties. The NeuroEvolution of Augmenting Topologies (NEAT) algorithm [16] is one of the most prominent NE methods due to its flexibility and capabilities to evolve competitive and compact networks [17]. NEAT uses a Genetic Algorithm (GA). Therefore, it uses two variation operators: recombination, which improves exploitation by sharing characteristics from two parents, and mutation, which enhances the exploration of the search space. NEAT uses five mutations: create or delete a connection between two nodes (neurons), create or delete a node, and replace a weight value. Moreover, NEAT implements a mechanism of speciation, which groups similar solutions to protect newly created individuals from other, more exploited solutions.

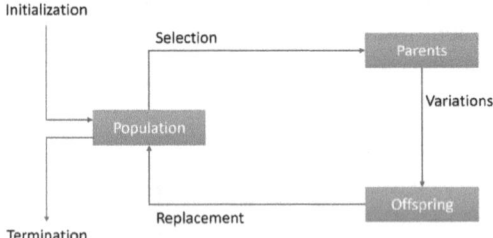

Fig. 4. A general schematic of an EA. The initialization forms the first individuals (solutions) of the population. A selection of individuals will form a subset of parents from which new individuals will be generated through variations (crossover and mutation). Afterward, the offspring will replace some individuals, forming the new population. This process is repeated until a termination criterion is reached.

The primary objective of this paper is to demonstrate the potential of the NEAT algorithm as an NE method in evolving SNNs capable of predicting epileptic seizures using EEG signals. The significance of this research lies in its potential to significantly improve the accuracy and efficiency of seizure prediction. The structure of this paper is as follows: Sect. 2 outlines the problem statement and the metrics that will be used. Section 3 provides a comprehensive literature review of AI methods for predicting epileptic seizures using EEG signals. Section 4 details the dataset and methodology used. Section 5 presents the experiment conducted, its results, and its observations. Finally, Sect. 6 offers the conclusions drawn from the research and outlines potential future work in this area.

2   Problem Statement

In this work, the problem of interest is the prediction of epileptic seizures using EEG data. The ability to categorize epileptic brain signals and the transition from interictal to pre-ictal stage, which marks the onset of a seizure [5], makes EEG data a promising input for this task. Notably, the problem is complex and highly temporal, necessitating a model that can effectively handle temporal data, such as SNNs. We will leverage the NEAT algorithm to address the complexity of designing and training SNNs.

To measure the performance of this method, three metrics will be used. The sensibility and specificity will show the model's capability to correctly predict true and false epileptic seizures. These metrics are shown in Eq. 3 and 4, where TP means True Positives and TN means True Negative predictions, whereas FN and FP mean False Negative and False Positive cases, respectively. On the other hand, the Seizure Prediction Horizon (SPH) will measure the prediction of the onset ($T_{detection}$) against the actual beginning of the episode (T_{actual}). Therefore, the higher the SHP, the better the ability to anticipate a seizure. This metric is shown in Eq. 5. The value of sensibility was chosen as the objective function to guide the evolutionary process. Therefore, the procedure is a maximization problem.

$$Sensibility = \frac{TP}{TP + FN} \tag{3}$$

$$Specificity = \frac{TN}{TN + FP} \tag{4}$$

$$SPH = T_{actual} - T_{detection} \tag{5}$$

The CHB-MIT [18] dataset will be used to test the proposed model. The data consists of scalp EEG signals (not invasive) of 24 patients at a 256 Hz sampling rate. From all records, 18 channels are consistently present following the 10–20 system (Fig. 1). Table 1 shows these channels and their corresponding signals. In total, 983 h of EEG activity are stored in different files. All records with ictal activity were annotated by experts at the beginning and end of this phase through visual inspection. Finally, 198 epileptic events were identified.

Table 1. Channel correspondence with the respective signal.

Channel	Signal	Channel	Signal
0	FP1 - F7	9	F4 - C4
1	F7 - T7	10	C4 - P4
2	T7 - P7	11	P4 - O2
3	P7 - O1	12	FP2 - F8
4	FP1 - F3	13	F8 - T8
5	F3 - C3	14	T8 - P8
6	C3 - P3	15	P8 - O2
7	P3 - O1	16	FZ - CZ
8	FP2 - F4	17	CZ - PZ

3   Related Work

The literature review shows several techniques to predict epileptic seizures using the CHB-MIT [18] database. Research has focused on the preprocessing of the EEG signals to extract helpful information between the regions of interictal and preictal. To obtain frequency and time domain characteristics, the authors of [19] used the Fast Fourier Transform (FFT). Likewise, Researches of [20, 21] performed the Short Time Fourier Transform (STFT). Moreover, in [22], Fourier coefficients (FC) were obtained to the same end. Obtaining Probability Density Functions (PDF) on different time windows was another approach to identifying interictal and preictal attributes proposed in [23]. A promising perspective is to use the entropy notion of signals. In [24], the authors used multivariate entropy, called Multivariate Modified Multiscale Distribution Entropy (MM-mDistEn), on all channels. Similarly, authors of [25] combined Transfer Entropy (TE) with Phase Transfer Entropy (PTE). Using Butterworth Filter (BwF) [26] and Wavelet Transform (WT) [27, 28] are other approaches that have been explored. Furthermore, other researchers have used Convolutional Neural Networks (CNN) to obtain preliminary features of the signals [21, 29, 30].

Different techniques have been explored to perform the classification and prediction of epileptic seizures after preprocessing. Researchers in [22, 23] have used a simple threshold process. Due to its high performance, authors [20, 27, 30] have applied Support Vector Machines (SVMs) to this task. ANNs have also been used for this purpose. Researchers in [24] explored Multi-layer Perceptron's (MLP) performance. Moreover, CNNs have also been used by [19, 21, 25, 26, 30, 31] in this aspect. Finally, due to its good performance on time-series applications, researchers [21, 28–30] have used Long Short-Term Memory (LSTM).

Table 2 summarizes all methods researched, including their metrics obtained. It can be seen that the most used method, both for preprocessing and prediction, is CNNs, even more than LSTMs. Nevertheless, no works have leveraged the properties of SNN for predicting epileptic seizures.

4   Methodology

This work aims to neuroevolved SNNs using NEAT capable of detecting the transition between interictal and pre-ictal phases using the 18 channels of CHB-MIT 1 as inputs. The database has two types of records: with and without ictal activity. The latter is characterized as having only interictal activity. Records with ictal activity have all four phases (Fig. 2). Nevertheless, the number of ictal activity records is disproportionate, with more instances of non-ictal records. To overcome this, 120 instances of non-ictal and ictal activity were generated. The ictal cases were segmented to have only continuous segments of interictal and pre-ictal, with a duration between 60 to 90 min [30]. Similarly, non-ictal cases were segmented to have only interictal activity, with a duration of 90 min. This approach allows us to detect the transition of interictal and pre-ictal phases.

One key aspect of SNNs is how they work and transmit information. Spiking neurons only produce single, discrete pulses, and connections between neurons

Table 2. Summary of proposed models for predicting epileptic seizures, sorted by year. Values in bold marked the best results. Entries with "-" specify data that does not apply.

Reference	Year	Methods	Sensibility	Specificity	SPH
[22]	2017	FC, Threshold	86.67%	86.67%	**45.30 min**
[19]	2019	FFT, CNN	91.50%	79.50%	5.00 min
[29]	2019	CNN, LSTM	91.88%	86.13%	21.00 min
[23]	2019	PDF, Threshold	90.30%	85.20%	22.63 min
[20]	2020	STFT, CNN, SVM	92.70%	90.80%	21.00 min
[30]	2021	CNN, LSTM	93.00%	92.50%	32.00 min
[24]	2021	MM-mDistEn, MLP	91.82%	**99.11%**	26.73 min
[21]	2022	STFT, CNN, LSTM	93.80%	91.20%	19.50 min
[25]	2022	TE-PTE, CNN	**99.52%**	97.47%	22.60 min
[27]*	2022	WT, SVM	-	-	30.00 min
[26]	2023	BwF, CNN	86.81%	86.45%	15.00 min
[28]	2023	WT, LSTM	90.72%	91.35%	30.00 min
[31]	2023	CNN	88.10%	92.10%	35.00 min

* [27] reported only an accuracy of 93.33%.

convey spike trains. Therefore, converting real-valued signals, such as EEG signals, into spike trains is essential. Following the work of [32], all EEG channel's signals were encoded to spike trains using Ben's Spiker Algorithm (BSA) optimized by a Differential Evolution (DE) approach. Finally, the dataset was partitioned into two segments: 70% for evaluating the performance of all individuals during the NE (Training set) and 30% for evaluating the best individual after NE (Validation set).

The NE process used the NEAT algorithm to evolve SNNs, with hyperparameters listed in Table 3. These were established trhough trial-and-error on preliminary executions with a reduced dataset, favoring low recombination and prioritizing adding connections/nodes over deleting them. Weight modifications were also favored. All SNNs were initialized with 18 inputs to use all EEG channels (Fig. 1). Both positive and negative connections were allowed. The evaluation of sensibility on the training set was set as the objective function. After the NE, the best individual obtained was evaluated using the validation set. Furthermore, the sensibility, specificity, and SPH metrics were reported. Five executions were performed to assess the method's consistency. All were carried out on an Intel Core i5-10300H CPU with 8 GB RAM and programmed in Python 3.9.

5   Experimental Results and Discussion

This work used the CHB-MIT database to predict epileptic seizures using SNNs produced by an NE approach. As mentioned in Sect. 4, the sensibility (Eq. 3)

Table 3. Parameter configuration for NEAT.

Hyperparameter		Value
General	Population	30
	Generations	400
	Max stagnation	3 gen
Recombination	Elitism	1 per species
	Umbral	40%
Mutation	Add connection	60%
	Delete connection	40%
	Add node	60%
	Delete node	20%
	Replace weight	80%
Speciation	Elitism	1
	Min size	1
	Similitude factor	2.4

was used as the objective function to guide the NE process on the training test. Afterward, the validation set was tested on the best individual obtained by the evolutionary process. In this last step, the specificity (Eq. 4) and SPH (Eq. 5) were also calculated.

The results of the NE method on all five executions showed a best performance of 94.12%, a median of 88.00%, and 88.14% ± 3.99 as mean and standard deviation. Furthermore, the mean of the number of neurons in all SNNs was 25.2 ± 4.01. The convergence plot of the median run is shown in Fig. 5. A growing trend is appreciated in it. Nevertheless, the behavior seems unsteady since NEAT applies a stagnation mechanism on complete species, deleting groups of individuals.

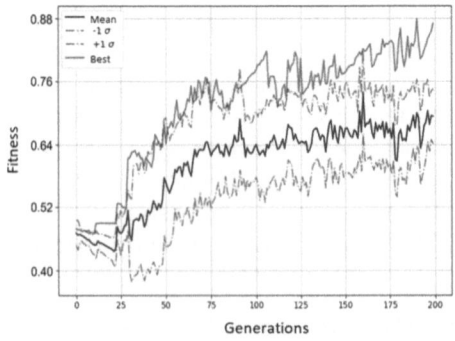

Fig. 5. Convergence plot of the median execution. The values of mean (blue), best (red), and variations (green) are depicted. (Color figure online)

Finally, Table 4 shows the validation results against the reviewed methods. As can be seen, the proposed method achieved a competitive sensibility value, attaining statistical metrics within the State-Of-The-Art (SOTA) range. The same can be argued with the specificity. Nevertheless, the gap between the best method [24] and the statistic reported is more considerable in the specificity. Finally, the method's best and median executions surpassed the SPH metric of the SOTA methods while achieving a competitive mean value. This fact highlights the ability of SNNs to process and handle temporal information. By integrating neurons with greater computational capacity in the temporal domain, SNNs performed better in extending the SPH.

Table 4. Comparison among the methods reviewed and the results on the validation set. Values in bold mean the best value.

Reference		Sensibility	Specificity	SPH
[22]		86.67%	86.67%	45.30 min
[19]		91.50%	79.50%	5.00 min
[29]		91.88%	86.13%	21.00 min
[23]		90.30%	85.20%	22.63 min
[20]		92.70%	90.80%	21.00 min
[30]		93.00%	92.50%	32.00 min
[24]		91.82%	**99.11%**	26.73 min
[21]		93.80%	91.20%	19.50 min
[25]		**99.52%**	97.47%	22.60 min
[26]		86.81%	86.45%	15.00 min
[28]		90.72%	91.35%	30.00 min
[31]		88.10%	92.10%	35.00 min
This work	Best	94.20%	88.57%	**50.34 min**
	Median	90.02%	79.63%	48.10 min
	Mean/σ	90.39% ± 3.11	81.65% ± 5.52	44.73 min ± 8.21

6 Conclusions and Future Work

Epilepsy is a worldwide common neurological condition where patients suffer from recurrent seizures. These episodes are defined as excessive electrical activity in brain regions, which can result in involuntary movement or even loss of consciousness. Furthermore, methods that monitor brain signals are often used to analyze these neurological disorders. Such is the case of EEG signals. To study this condition, researchers have identified 4 phases of epileptic brain signals: interictal, pre-ictal, ictal, and post-ictal, representing different stages of an episode. The progression of the interictal to the pre-ictal ensures the onset of a seizure.

This work proposed using the NEAT algorithm, an NE method, to evolve SNNs capable of predicting epileptic seizures by analyzing the transition between interictal and pre-ictal stages. The CHB-MIT database assessed the model's performance, measuring sensibility, specificity, and SPH. The first two indicate the ability to detect true positive and false positive epileptic seizure predictions, whereas the latter estimate the ability to forecast the onset of an episode.

The results show a competitive performance compared with SOTA methods, achieving a mean value of 94.39% sensibility and 81.65% specificity. Interestingly, the model increased the SPH reported, reaching 48.73 min of anticipation in the best result and 48.19 min in the median. The mean value was in range with SOTA results, just under 0.57 min short of the best value reported up to date. Furthermore, the mean of the number of neurons in the SNNs evolved was 25.2. This analysis reinforces the properties that arise using neural models as computational units. SNNs excel at time-related problems by adding more powerful temporal neurons. Furthermore, it was shown that the proposed method is functional and promising in predicting epileptic seizures. By increasing the time of anticipation, patients can be prevented, and further actions might be efficiently applied, such as administering anti-seizure medication, alerting family members, or finding a safe place, thereby reducing the risks of accidents [33].

Nevertheless, more experimentation is needed. In this work, only five executions could be reported due to the computational time required. Different approaches, such as surrogate methods, could reduce this restriction. Other objective functions could be explored. An interesting approach might be combining the SPH and sensibility to assess the possibility of increasing the forecast value without compromising effective prediction. Additionally, since the CHB-MIT database contains only 24 patients, future work could consider artificial instance generation techniques to enhance model robustness.

Acknowledgments. The first (CVU 1075919) and the fourth (CVU 1142850) authors acknowledge support from the Mexican National Council for Humanities, Science and Technology (CONAHCyT) through a scholarship to pursue postgraduate studies at the University of Veracruz.

References

1. World Health Organization. (2024). Epilepsy. World Health Organization. https://www.who.int/news-room/fact-sheets/detail/epilepsy
2. Jennum, P., Gyllenborg, J., Kjellberg, J.: The social and economic consequences of epilepsy: a controlled national study. Epilepsia **52**, 949–956 (2011). https://doi.org/10.1111/j.1528-1167.2010.02946.x
3. Jirsa, V.K., Stacey, W.C., Quilichini, P.P., Ivanov, A.I., Bernard, C.: On the nature of seizure dynamics. Brain J. Neurol. **137**(Pt 8), 2210–2230 (2014). https://doi.org/10.1093/brain/awu133

4. St. Louis, E.K., et al. (eds.).: Electroencephalography (EEG): An Introductory Text and Atlas of Normal and Abnormal Findings in Adults, Children, and Infants. American Epilepsy Society (2016)
5. An, S., Kang, C., Lee, H.W.: Artificial intelligence and computational approaches for epilepsy. J. Epilepsy Res. **10**(1), 8–17 (2020). https://doi.org/10.14581/jer.20003
6. Dissanayake, T., Fernando, T., Denman, S., Sridharan, S., Fookes, C.: Patient-independent epileptic seizure prediction using deep learning models (2020). arXiv preprint arXiv:2011.09581
7. Shoeibi, A., et al.: Epileptic seizures detection using deep learning techniques: a review. Int. J. Environ. Res. Public Health **18**(11), 5780 (2021). https://doi.org/10.3390/ijerph18115780
8. Maass, W.: Networks of spiking neurons: the third generation of neural network models. Neural Netw. **10**(9), 1659–1671 (1997). https://doi.org/10.1016/S0893-6080(97)00011-7
9. Tavanaei, A., Ghodrati, M., Kheradpisheh, S.R., Masquelier, T., Maida, A.: Deep learning in spiking neural networks. Neural Netw. **111**, 47–63 (2019). https://doi.org/10.1016/j.neunet.2018.12.002
10. Wang, G., et al.: Brain-inspired artificial intelligence research: a review. Sci. China Technol. Sci. 1–15 (2024). https://doi.org/10.1007/s11431-024-2732-9
11. Ponulak, F., Kasinski, A.: Introduction to spiking neural networks: information processing, learning and applications. Acta Neurobiologiae Experimentalis **71**(4), 409–433 (2011). https://doi.org/10.55782/ane-2011-1862
12. Stein, R.B.: Some models of neuronal variability. Biophys. J. **7**(1), 37–68 (1967). https://doi.org/10.1016/S0006-3495(67)86574-3
13. Gerstner, W., Kistler, W.M., Naud, R., Paninski, L.: Neuronal Dynamics: From Single Neurons to Networks and Models of Cognition. Cambridge University Press, Cambridge (2014)
14. Neftci, E.O., Mostafa, H., Zenke, F.: Surrogate gradient learning in spiking neural networks: bringing the power of gradient-based optimization to spiking neural networks. IEEE Signal Process. Mag. **36**(6), 51–63 (2019). https://doi.org/10.1109/MSP.2019.2931595
15. Taherkhani, A., Belatreche, A., Li, Y., Cosma, G., Maguire, L.P., McGinnity, T.M.: A review of learning in biologically plausible spiking neural networks. Neural Netw. **122**, 253–272 (2020). https://doi.org/10.1016/j.neunet.2019.09.036
16. Stanley, K.O., Miikkulainen, R.: Evolving neural networks through augmenting topologies. Evolut. Comput. **10**(2), 99–127 (2002). https://doi.org/10.1162/106365602320169811
17. Stanley, K.O., Miikkulainen, R.: Efficient evolution of neural network topologies. In: Proceedings of the 2002 Congress on Evolutionary Computation. CEC'02 (Cat. No. 02TH8600), vol. 2, pp. 1757–1762. IEEE (2002, May). https://doi.org/10.1109/CEC.2002.1004508
18. Shoeb, A.H.: Application of machine learning to epileptic seizure onset detection and treatment (Doctoral dissertation, Massachusetts Institute of Technology) (2009)
19. Liu, C.L., Xiao, B., Hsaio, W.H., Tseng, V.S.: Epileptic seizure prediction with multi-view convolutional neural networks. IEEE Access **7**, 170352–170361 (2019). https://doi.org/10.1109/ACCESS.2019.2955285
20. Usman, S.M., Khalid, S., Aslam, M.H.: Epileptic seizures prediction using deep learning techniques. IEEE Access **8**, 39998–40007 (2020). https://doi.org/10.1109/ACCESS.2020.2976866

21. Aslam, M.H., et al.: Classification of EEG signals for prediction of epileptic seizures. Appl. Sci. **12**(14), 7251 (2022). https://doi.org/10.3390/app12147251

22. Chu, H., Chung, C.K., Jeong, W., Cho, K.H.: Predicting epileptic seizures from scalp EEG based on attractor state analysis. Comput. Methods Programs Biomed. **143**, 75–87 (2017). https://doi.org/10.1016/j.cmpb.2017.03.002

23. Ibrahim, F., et al.: A statistical framework for EEG channel selection and seizure prediction on mobile. Int. J. Speech Technol. **22**(1), 191–203 (2019). https://doi.org/10.1007/s10772-018-09565-7

24. Aung, S.T., Wongsawat, Y.: Prediction of epileptic seizures based on multivariate multiscale modified-distribution entropy. PeerJ. Comput. Sci. **7**, e744 (2021). https://doi.org/10.7717/peerj-cs.744

25. Mu, Y., Zhang, X., Zhang, M., Wang, H.: Epilepsy prediction based on PTE and TE of EEG signals using DSC-CNN. In 2022 12th International Conference on Information Science and Technology (ICIST), pp. 193–200. IEEE (2022). https://doi.org/10.1109/ICIST55546.2022.9926907

26. Ding, X., Nie, W., Liu, X., Wang, X., Yuan, Q.: Compact convolutional neural network with multi-headed attention mechanism for seizure prediction. Int. J. Neural Syst. **33**(3), 2350014 (2023). https://doi.org/10.1142/S0129065723500144

27. Perez-Sanchez, A.V., Valtierra-Rodriguez, M., Perez-Ramirez, C.A., De-Santiago-Perez, J.J., Amezquita-Sanchez, J.P.: Epileptic seizure prediction using wavelet transform, fractal dimension, support vector machine, and EEG signals. Fractals **30**(07), 2250154 (2022). https://doi.org/10.1142/S0218348X22501547

28. Pandey, A., Singh, S.K., Udmale, S.S., Shukla, K.K.: An intelligent optimized deep learning model to achieve early prediction of epileptic seizures. Biomed. Signal Process. Control **84**, 104798 (2023). https://doi.org/10.1016/j.bspc.2023.104798

29. Wei, X., Zhou, L., Zhang, Z., Chen, Z., Zhou, Y.: Early prediction of epileptic seizures using a long-term recurrent convolutional network. J. Neurosci. Methods **327**, 108395 (2019). https://doi.org/10.1016/j.jneumeth.2019.108395

30. Usman, S.M., Khalid, S., Bashir, Z.: Epileptic seizure prediction using scalp electroencephalogram signals. Biocybern. Biomed. Eng. **41**(1), 211–220 (2021). https://doi.org/10.1016/j.bbe.2021.01.001

31. Ji, H., et al.: An effective fusion model for seizure prediction: GAMRNN. Front. Neurosci. **17**, 1246995 (2023). https://doi.org/10.3389/fnins.2023.1246995

32. López-Herrera, C.A., Acosta-Mesa, H.G., Mezura-Montes, E.: A surrogate-assisted differential evolution approach for the optimization of ben's spiker algorithm parameters. In: Mexican International Conference on Artificial Intelligence, pp. 337–348. Springer Nature, Switzerland, Cham (2023)

33. Ramgopal, S., et al.: Seizure detection, seizure prediction, and closed-loop warning systems in epilepsy. Epilepsy Behav. E&B **37**, 291–307 (2014). https://doi.org/10.1016/j.yebeh.2014.06.023

Identification of Simple Geometric Figures Using Matlab and ROS

Atalia-Yael Hernández-Sánchez$^{(\boxtimes)}$ (ID), Jorge Aramburo-Aguilar(ID),
Héctor-Gabriel Acosta-Mesa(ID), Sergio Hernandez-Mendez(ID),
and Antonio Marin-Hernandez(ID)

Artificial Intelligence Research Institute, University of Veracruz,
Campus Sur, Calle Paseo No 112, 91097 Xalapa, Veracruz, México
{zS23000639,zS23000638}@estudiantes.uv.mx,
{heacosta,sergihernandez,anmarin}@uv.mx

Abstract. Service robotics must balance the design and development of functional units with simple human-robot communication to generalize its use and service to the public. This paper presents an implementation that combines color and shape detection to recognize four figures with specific shapes and colors. Each figure represents a basic control instruction that will be sent to a differential drive robot. The algorithm uses images extracted from a webcam using MATLAB computer vision libraries, and then processed by methods such as binary mask generation, edge detection with a Canny filter, edge dilation, and detection of straight lines with the Standard Hough Transform. At the same time, the instructions are executed in the robot model with the help of packages developed with ROS approaching real-time control. Tests were carried out in a semi-controlled environment with ambiental disturbances, two different lighting conditions, and digital images to validate our proposal.

Keywords: Simultaneous features detection · Service robot ·
Differential drive robot model

1 Introduction

The notable and continuous advancement of technology has allowed humans to interact with intelligent machines in practically every aspect of their lives. Given the indisputable advances in artificial intelligence used in practically all fields knowledge, multiple computer vision methods have emerged that complement the branch of robotics.

Due to the hardware improvement and the increase in processing and storage capabilities seen since 2010, robot vision has obtained its identity. This area is born from the necessity of imitating the sense of sight. Since, human beings depend to a significant degree on it to interpret and act according to what their environment presents to them. Therefore, it is based on techniques studied in areas such as computer vision and machine learning [4].

L. Martínez-Villaseñor et al. (Eds.): MICAI 2024 Workshops, LNAI 15465, pp. 147–154, 2025.
https://doi.org/10.1007/978-3-031-83882-8_14

Unlike computer vision, robot vision considers not only the information extraction from given images or videos but also real-time data that will influence the robot's future actions. Thus, generating a sequence of observation, interpretation, and interactive action. Bringing with it challenges begins with analyzing how the robot receives environmental stimuli. These systems frequently rely on signals transmitted from mobile sensors, instead of specific data or images from fixed cameras. Producing a set of issues such as orientation problems, noise caused by lighting and movement present in the environment, and object occlusion, in addition to others. We can group these setbacks into eight categories: active vision, anomaly detection, interest detection, semantic scene understanding, place recognition, Simultaneous Localization and Mapping (SLAM), vision-based control, and the development of efficient and real-time solutions for relevant application areas [4].

In the process of designing a robot, the implementation of a human-machine communication system that is user-friendly for inexperienced is not always given priority in management. Consequently, it is necessary to explore avenues of approach beyond the use of coded instructions and consider the implementation of graphical aids instead.

This paper proposes a color detection algorithm complemented with a shape detector to provide essential navigation commands to a service robot. The goal of this is to realize real-time communication between a service robot and its human operator. Employed in this context uncomplicated visual elements, combining color and shape features to avoid reliance on immediate instruction coding or user language.

The following four sections will detail the selected approach. The second part presents projects focused on color and shape detection to establish the robot vision's approaches and uses. Section three details the theory followed to implement the proposal described in the introduction. Empirical results are summarized in headland four. Finally, conclusions and future work are summarized.

2   Background

Robot design and control are complex tasks due to the need to develop and combine both software and hardware. Which consists of the entire mechanical framework, including sensors, actuators, processing logic, and control schemes. An effective way to develop them is through a unified and open platform that allows hardware and software developers to work together such as the Robot Operating System (ROS)[1]. ROS is a flexible framework that relies on a host operating system, which is usually Ubuntu, to write software and simplify the creation and simulation of robots on various platforms. This platform has achieved significant popularity among robot developers worldwide, attributed to its rich repository of open-source libraries and tools [16].

---

[1] Project focused on designing and implementing a platform for the development of robots [17].

Besides, MATLAB is a powerful tool focused on the design and analysis of control systems whose integrated graphics facilitate visualization and information extraction. MATLAB has powerful data processing, a great capacity for visual drawing, many functions, and its environment for algorithm development. Also, the Robotics System Toolbox can be added, incorporating a set of tools that allows users to interact with ROS and use ROS functionalities in MATLAB [16].

Object tracking is a crucial application of computer vision, as it requires addressing the challenges of illumination and occlusion. There are proposals for a robust algorithm based on colors analyzed in the HSV space to be used to track objects and deal with occlusion. In brief, by extracting the frames of one or several videos, their RGB values can be converted to their corresponding HSV parameters. Those parameters are used to threshold the images according to an HSV range and, after applying morphological operations, obtain the area of the objects of interest. With these areas, it is easy to contour the objects and, track them even when other dynamic elements of the environment occlude them [5]. A different approach to color detection would be to use the histogram of oriented gradient (HOG) mixed with color features. This method shows good results with diverse shapes such as fruits [8].

A robotic arm with a gripper can use computer vision-based algorithms to select and arrange objects. Setting apart several objects according to their colors, geometrical shapes, and sizes. Extracting color data from a PixyCMU camera sensor, and processing the shape and size detection by OpenCV [1]. For edge detection, mixed methods can also be found that combine two different detection algorithms and complement each other. An example of this is the fusion of the improved polarimetric constant false alarm rate (IP_CFAR) edge detector and the proposed weighted gradient-based (WG) [15]. One method that shows a different approach to color and shape detection is that of Fahad K., which integrates these two detections by focusing on color characteristics [7].

A recent instance showcases the importance of robot vision in the domain of autonomous vehicle control systems, such as automobiles. These vehicles need to detect and recognize traffic lights. This predicament has already been solved with an algorithm that fuses fast radial symmetry transformation, Hough Transformation, and color segmentation. The preliminary placement of light areas is determined based on their color, followed by the specification of their shape and position. Lastly, Hough Transform is employed to refine the results [14].

3   Color and Shape Detection Method

The above-mentioned proposal was pursued through a method aligned with the previously detailed robot vision concept. Employing color and shape recognition techniques using MATLAB, concurrently with the implementation of a differential robot model's simulation and control via ROS and Gazebo. The summary of the workflow can be observed in Fig. 1.

Taking advantage of the Artificial Intelligence Research Institute's resources, we used a Pioneer 3D-X based differential robot model known by the Institute

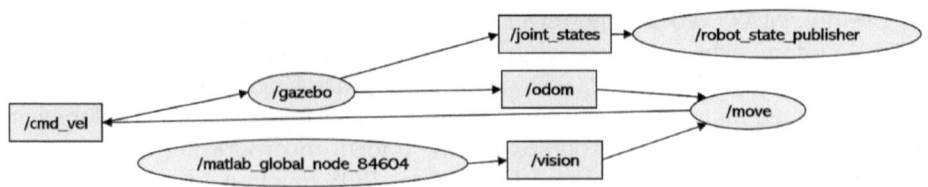

Fig. 1. Graph of active processes in ROS during program execution.

such as UVerto. Said model was developed on ROS Noetic, therefore including a simulated world built in Gazebo and several simulation tools that emulate real components. Some of them are its RGBD camera, laser, odometry and AMCL packages, and so on. The Institute has its server running on Linux to work out UVerto and its virtual model, though UVerto's model is stored in a GitHub repository too.

Because UVerto is being engineered as a service robot, a graphic system of communication was considered appropriate to implement, given that specialized programming is necessary for most of its control.

3.1   Image Acquisition and Processing

The first workflow phase envelops computer vision methods implemented on MATLAB and complemented with a webcam integrated into a laptop. In addition two toolboxes: MATLAB Support Package for USB Webcams [12] and ROS Toolbox. [10]. The first one enables communication between MATLAB and webcams, whereas the second one connects MATLAB, Simulink, and ROS.

Considering MATLAB's frames extraction from webcam video [11], in this case, with a 30 FPS average frequency, the algorithm requires a loop to image acquisition and processing operating during the webcam's use, inching nearer to real-time data processing. Hence all described next steps are repeated for each frame. Once an image corresponding to a video frame is obtained, previously calculated calibration parameters are used to reduce the lens distortion [18].

Keeping the RGB image type obtained in mind, each color channel is extracted individually [6] and subtracted from the gray-scale original image. Subsequently, heuristic thresholds are established to generate binary masks corresponding to color segmentation.

Afterward, a Canny filter [2] detects edges in each mask, and traced borders are dilated helping line detection. Straight lines are searched with the Standard Hough Transform [9] and, if they match the color and number of sides of each figure, a string-type message is published to indicate which figure is being viewed. Altogether the algorithm can detect four figures: red triangle, blue square, green pentagon, and purple star. All of them were digitally generated to determine exactly their RGB value.

3.2   Model Control Using ROS

The differential drive consists of two wheels assembled on a single axe controlled by two independent motors, one for each wheel. The wheel's velocity determines the robot's movement. The following wheel's relative velocity variation will be the variation of the rotation point and, therefore, of the path that the robot will follow [3].

Consequently, considering UVerto's model such as a differential robot, we can determine its possible position and orientation according to two control parameters: velocities along the ground at the left tire vl and right tire vr. This problem is known as such differential drive kinematics and can be solved with vl and vr parameters, the unsigned distance from ICC to the midpoint between the two wheels R, and the angular velocity ω given as functions in time. All things being considered is easy to show that if the robot has the position (x, y, θ) at a time t, then at the instant $t + \delta t$ the pose of the robot is denoted by:

$$\begin{bmatrix} x' \\ y' \\ \theta' \end{bmatrix} = \begin{bmatrix} cos(\omega\delta t) & -sin(\omega\delta t) & 0 \\ sin(\omega\delta t) & cos(\omega\delta t) & 0 \\ 0 & 0 & 1 \end{bmatrix} * \begin{bmatrix} x - ICC_x \\ y - ICC_y \\ \theta \end{bmatrix} + \begin{bmatrix} ICC_x \\ ICC_y \\ \omega\delta t \end{bmatrix}. \tag{1}$$

Essentially: Eq. (1) describes the movement of a robot as it rotates a distance R on its ICC at a given angular velocity ω. By integrating this equation with an initial condition (x_0, y_0, θ_0) it is possible to calculate the robot's position at any given time t based on the control parameters $v_l(t)$ and $v_r(t)$ [3].

In contrast, forward kinematics equations can be used to select the control parameters and set specific trajectories and global poses for the robot. For this perspective, it is important to give due consideration to the two specific cases of differential vehicle motion. If $v_l = v_r = v$, the robot moves in a straight line, i.e.:

$$\begin{bmatrix} x' \\ y' \\ \theta' \end{bmatrix} = \begin{bmatrix} x + vcos(\theta)\delta t \\ y + vsin(\theta)\delta t \\ \theta \end{bmatrix}. \tag{2}$$

Given that $-v_l = v_r = v$, the following statement is warranted:

$$\begin{bmatrix} x' \\ y' \\ \theta' \end{bmatrix} = \begin{bmatrix} x \\ y \\ \theta + 2v\delta t/l \end{bmatrix}, \tag{3}$$

which indicates that the robot is rotating in place, without going forward [3].

For the current assignment, each figure's necessary time on camera was calculated using Eqs. (2) and (3) to move and rotate the robot to approximate desired distances. For example, if the robot must go straight forward 1 m onto the x axis from the origin, the initial condition $(x_0 = 0, y_0 = 0, \theta_0 = 0)$ and a constant velocity $v = v_l = v_r = 0.2 m/s$ are considered to reach a final position final $(x_1 = 1, y_1 = 0, \theta_1 = 0)$. Considering this given information and Eq. (2), it is obtained $dt = 5s$. Similarly, the required rotating times were calculated.

When a message is published, the link-up between MATLAB and ROS is created, since any subscriber topic codified in ROS can catch said message [13]. Based on this, a subscriber package was generated to subscribe to the message published by MATLAB and to the odometry topic of UVerto's simulation. Next, it publishes a message to the UVerto's cmd_vel topic to set linear and angular velocities. In this manner, the model can move around its environment according to what the laptop's camera observes.

4   Results

The four essential movements and their corresponding figures are: red triangle indicates that the robot must go straight ahead, blue square sends a message to turn counterclockwise the robot, green pentagon means that the robot will turn clockwise, and purple star stops all movements. The tests consisted of showing the figures to the webcam until they were detected by the algorithm, holding them for five seconds in the visible range, and finally assessing the successes and errors in detecting. All of them were performed in a bedroom with a window such light source, pointing the camera towards the window, and positioning the images in front of the webcam. In addition, external light sources were avoided as far as possible.

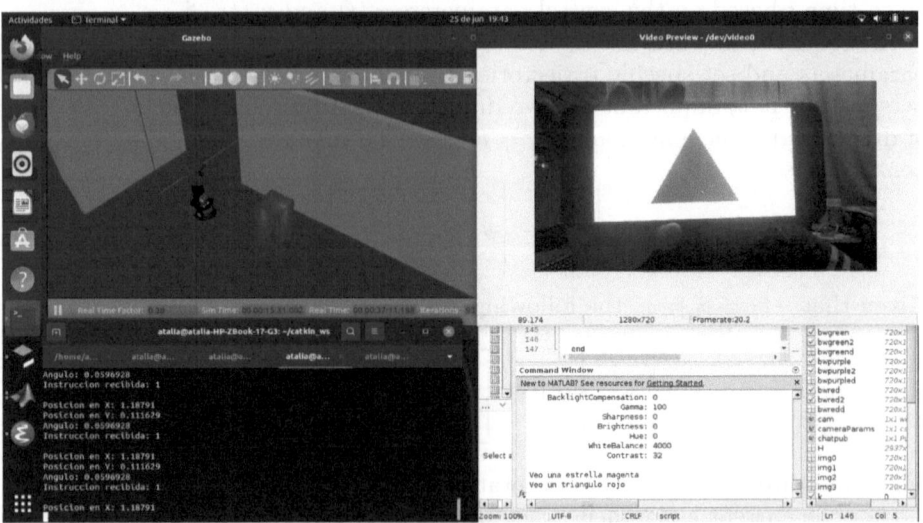

Fig. 2. Implementation of the algorithm in the simulator.

As is shown in Fig. 2, UVerto's simulation is running in Gazebo and its movement is caused by detecting a red triangle on the camera from MATLAB.

Table 1. Comparison of accuracy between the tests performed at 7:00hrs and 17:30hrs.

	Hour - 7:00				Hour - 17:30			
Figure	Test 1	Test 2	Test 3	Test 4	Test 1	Test 2	Test 3	Test 4
Red triangle	11 of 12	11 of 13	13 of 14	8 of 10	17 of 17	14 of 14	15 of 17	18 of 18
Blue square	7 of 10	8 of 9	12 of 14	8 of 10	13 of 17	13 of 19	15 of 15	15 of 15
Green pentagon	9 of 13	8 of 12	11 of 12	9 of 11	14 of 14	16 of 16	16 of 16	17 of 17
Purple star	7 of 7	9 of 9	7 of 7	10 of 10	15 of 15	15 of 17	12 of 12	18 of 18
Total	34 of 42	36 of 43	43 of 47	35 of 41	59 of 63	58 of 60	58 of 60	68 of 68
Average accuracy	80.96%	83.73%	91.49%	85.37%	93.66%	97.63%	96.67%	100%
Average accuracy per hour	85.38%				96.99%			

The Table 1 presents the results of tests conducted with various figures at different times (specifically, midday and sunset), aimed at observing the impact of lighting on the accuracy of the algorithm.

We can observe that the accuracy decreases as the ambient lighting increases, due to the use of an electronic device to display the figures.

5   Conclusions and Discussion

In agreement with all computer vision projects, image quality challenges were presented, which resulted in a significant obstacle during the segmentation color thresholding. Applying two conditions to determine figures broadens the scope of detection compared to methods that only focus on detecting single-colored circles or objects. This approach enables the detection of objects with the same color but different shapes, and vice versa. The average accuracy of the algorithm is 92.94%, in an semi-controlled environment that depends on natural lighting and does not have as many restrictions as a laboratory. Future work includes: modifying, using, and creating packages and drivers to achieve more complex control of the robot; changing the color space to HSI to deal with lighting issues and facilitate the use of complex colors; and refining shape detection to recognize more elaborate shapes in different orientations more quickly than the current one.

References

1. Abdullah-Al-Noman, M., Eva, A., Yeahyea, T., Khan, R.: Computer vision-based robotic arm for object color, shape, and size detection. J. Robot. Control (JRC) **3**(2), 180–186 (2022). https://doi.org/10.18196/jrc.v3i2.13906, https://journal.umy.ac.id/index.php/jrc/article/view/13906
2. Canny, J.: A computational approach to edge detection. IEEE Trans. Pattern Anal. Mach. Intell. **PAMI-8**(6), 679–698 (1986). https://doi.org/10.1109/TPAMI.1986.4767851
3. Dudek, G., Jenkin, M.: Computational Principles of Mobile Robotics. Cambridge University Press, Cambridge (2013)

4. van Eden, B., Rosman, B.: An overview of robot vision. In: 2019 Southern African Universities Power Engineering Conference/Robotics and Mechatronics/Pattern Recognition Association of South Africa (SAUPEC/RobMech/PRASA), pp. 98–104 (2019). https://doi.org/10.1109/RoboMech.2019.8704781

5. Gajbhiye, S.D., Gundewar, P.P.: A real-time color-based object tracking and occlusion handling using arm cortex-a7. In: 2015 Annual IEEE India Conference (INDICON), pp. 1–6 (2015). https://doi.org/10.1109/INDICON.2015.7443641

6. Gil, P., Torres, F., Ortiz Zamora, F.G.: Detección de objetos por segmentación multinivel combinada de espacios de color (2004)

7. Khan, F.S., Anwer, R.M., van de Weijer, J., Bagdanov, A.D., Vanrell, M., Lopez, A.M.: Color attributes for object detection. In: 2012 IEEE Conference on Computer Vision and Pattern Recognition, pp. 3306–3313 (2012). https://doi.org/10.1109/CVPR.2012.6248068

8. Liu, X., Zhao, D., Jia, W., Ji, W., Sun, Y.: A detection method for apple fruits based on color and shape features. IEEE Access **7**, 67923–67933 (2019). https://doi.org/10.1109/ACCESS.2019.2918313

9. MathWorks: hough.   https://la.mathworks.com/help/images/ref/hough.html?lang=en. Accessed 24 June 2024

10. MathWorks: Ros network access in matlab. https://la.mathworks.com/help/ros/ros-in-matlab.html. Accessed 24 June 2024

11. MathWorks: snapshot.   https://la.mathworks.com/help/matlab/supportpkg/webcam.snapshot.html. Accessed 24 June 2024

12. MathWorks: webcam.   https://la.mathworks.com/help/matlab/supportpkg/webcam.html. Accessed 24 June 2024

13. ROS.org: Comprendiendo tópicos ros. https://wiki.ros.org/es/ROS/Tutoriales/ComprendiendoTopicosROS. Accessed 24 June 2024

14. Shakirzyanov, R.M., Shakirzyanova, A.A.: Object detection using color segmentation, radial symmetry detector, and hough method. In: 2021 International Russian Automation Conference (RusAutoCon), pp. 714–718 (2021). https://doi.org/10.1109/RusAutoCon52004.2021.9537348

15. Shi, J., Jin, H., Xiao, Z.: A novel hybrid edge detection method for polarimetric sar images. IEEE Access **8**, 8974–8991 (2020). https://doi.org/10.1109/ACCESS.2020.2963989

16. Tang, W.J., Liu, Z.T.: A convenient method for tracking color-based object in living video based on ros and matlab/simulink. In: 2017 2nd International Conference on Advanced Robotics and Mechatronics (ICARM), pp. 724–727 (2017). https://doi.org/10.1109/ICARM.2017.8273251

17. Wyrobek, K.A., Berger, E.H., Van der Loos, H.M., Salisbury, J.K.: Towards a personal robotics development platform: rationale and design of an intrinsically safe personal robot. In: 2008 IEEE International Conference on Robotics and Automation, pp. 2165–2170 (2008). https://doi.org/10.1109/ROBOT.2008.4543527

18. Zhang, Z.: A flexible new technique for camera calibration. IEEE Trans. Pattern Anal. Mach. Intell. **22**(11), 1330–1334 (2000). https://doi.org/10.1109/34.888718

Experimental Study for Automatic Feature Construction to Segment Images of Lungs Affected by COVID-19 Using Genetic Programming

David Herrera-Sánchez$^{(\boxtimes)}$ ⓘ, José-Antonio Fuentes-Tomás ⓘ,
Héctor-Gabriel Acosta-Mesa ⓘ, and Efrén Mezura-Montes ⓘ

Artificial Intelligence Research Institute, University of Veracruz, 91097 Xalapa,
Veracruz, Mexico
hersan19@hotmail.es, {heacosta,emezura}@uv.mx

Abstract. Image segmentation is a challenging task due to image variations, such as illumination, background, noise, and others. There are several segmentation methods, but the requirement of prior knowledge and parameter setting makes it hard to perform a good segmentation, especially in medical images where an expert is needed to make the segmentation accordingly with the prior knowledge to determine the area of interest. In this work, we are focused on the feature construction to make an image segmentation of computerized tomography scans of lungs affected by COVID-19. Genetic Programming (GP) is used to evolve a program to extract and construct features from the image to make a segmentation where the target is to find the affected area. The flexibility that offers GP allows us to face the segmentation task and know which functions are used in the final program, leading to an interpretable solution. The results of the experiments demonstrate that GP is capable of extracting and constructing features from the Computerized Tomography images to perform the segmentation of lungs affected by COVID-19, achieving values of 0.59 of $F_1 - score$ metric to measure the segmentation performance. Furthermore, the experimental results determine the appropriate parameters for the evolutionary process.

Keywords: Genetic Programming · Image Segmentation · Image Feature Construction

1 Introduction

The coronavirus disease (COVID-19), which originated in Wuhan, China, was highly contagious and spread worldwide, causing different critical statuses in economics, transportation, and, most importantly, the healthcare system. The World Health Organization (WHO) declared an international pandemic in December 2019, which was terminated in May 2023 [1]. However, mutated strains

L. Martínez-Villaseñor et al. (Eds.): MICAI 2024 Workshops, LNAI 15465, pp. 155–166, 2025.
https://doi.org/10.1007/978-3-031-83882-8_15

have emerged in many countries at present. We already know that the most effective test to determine if you are infected by this virus is the Polymerase Chain Reaction (PCR) test [1]; the problem is the time this process takes compared with the evolution of the infection in the patients. In some cases, patients already have symptoms, and the worst cases already have complications in the respiratory system, especially in the lungs. Furthermore, PCR tests should be done carefully, and the sample must be transported under appropriate conditions. Due to the long time that the PCR test takes and the difficult task of transporting the sample, computerized systems help to carry out a pre-evaluation of the patient. Thus, the time to diagnose the patient is shortened to be treated, avoiding complications.

Medical imaging, such as X-ray and Computerized Tomography, is important in helping medical specialists. It is a tool that saves time and helps to make clinical decisions and diagnoses [7]. Fortunately, the scientific community has shared datasets of chest Computerized Tomography (CT) scans of COVID-19 cases. An expert segments these CT scans to determine the affected lung area. However, segmentation by hand is hard work and requires time and prior knowledge. Image segmentation is the process that separates an image into representative regions according to criteria that are shared in the region. Correct segmentation can improve the system's performance.

For our purpose, competitive segmentation can help improve diagnoses from medical specialists. Segmentation is an important step in the analysis and assessment of the area affected by COVID-19. Although the segmentation of medical images supported by GP is a little explored area, some works have been carried out on feature extraction and construction for image classification of skin lesions [2] and biomedical image segmentation identifying cells [9].

The proposal is to use GP to extract the most representative information to segment the affected lung area caused by COVID-19. In this paper, the use of GP is proposed to deal with the segmentation task and investigate the following objectives:

1. Propose a function to make a threshold for the image's pixel values.
2. Evolve a set of functions of basic operators and functions of image processing for grayscale images to find representative features to segment the CT images.
3. Comparing two segmentation methods supported by automatic feature construction and extraction from the images.

This paper is organized as follows: Sect. 2 introduces the basic concepts of GP, and the related work of GP for image processing. Section 3 describes the proposed approach. Section 4 contains the general experiment design. Section 5 shows the experiments, results and analysis. Finally, Sect. 6 presents the conclusion and future work.

2   Background and Related Work

2.1   Genetic Programming

Genetic programming (GP) is a paradigm of evolutionary computation (EC). Based on Darwin's evolutionary theory, EC works with a population in which each individual represents a potential solution for a problem. Within the population, individuals compete among themselves to survive. The fittest individuals are the most likely to survive evolution. Also, the evolution process generates new individuals through crossover or mutation operators [7]. The new individuals contain the parent's features, which it is expected that can solve the problem according to the features inherited. The evolution is performed until it reaches a known stop criterion. It can be set according to the number of generations, the number of evaluations, or with a low population diversity.

The difference between the paradigms within the EC is the representation. GP is well-known for its tree representation, which encourages interpretable solutions. Tree contain functions, variables, and constants. The internal nodes contain the functions used to solve the problem, such as mathematical operators or functions for image processing, as in our case. The leaf nodes contain constant values or variables used by the internal nodes.

One problem known within GP is the bloat effect. It occurs when, throughout crossover and mutation, the new individuals' complexity is increased without improving their fitness. A simple solution is to reject the offspring that exceeds a determined depth copying their parents. In [11], a maximum allowed depth value of 17 is suggested.

2.2   GP for Image Processing

GP has demonstrated robustness in dealing with several visual tasks, with image classification being the dominant task in *state-of-art* [2,7,8] in contrast with image segmentation [12,16,18].

GP systems for visual tasks usually consist of arithmetic and computer vision-related operators, such as filters or morphological operators [15] to extract and construct a new meaningful composition of features which are further processed by a classifier. Inspired by the deep learning approach, some works have included convolutional operators allowing the evolution of the filter's weights and size [3,5].

In contrast to image classification, image segmentation requires observing each pixel as a sample. Instead of using the pixels as inputs, Song and Ciesielski [16], propose texture classifiers employing fragments of images with several textures as inputs, reaching the identification of boundaries of different textures, even for those not regular, with a faster segmentation method than the conventional ones. They consider nine operations between conditionals, comparative, and arithmetic functions as the function set, and two terminals are the primitive set. Their results are evaluated in two datasets using the accuracy as the fitness function.

The methods of image segmentation and the evaluation measures depend on the image type (grayscale, color, textures) and its domain (medical, natural, or satellite). Several segmentation measures can evaluate only a few aspects of a segmented image, such as the entropy, variance, or homogeneity regions. To overcome these limitations, Vojodit *et al.* [18] constructed a new evaluation measure combining single measures with GP using four arithmetic operators as the function set and the single evaluation functions as terminals. In addition to the arithmetic functions, Liang *et al.* [12] employs filtering and thresholding operators, such as Otsu's method, totaling 14 functions. They propose three methods called *StronglyGP, TwoStageGP, and CoevoGP.* The *TwoStageGP* evolves preprocessing and postprocessing algorithms separately with two different evolutionary processes, where overfitting and an increment in consumption time were observed. In contrast, in *CoevoGP* use a single evolutionary process dividing the problem into i) the preprocessing and segmentation step and ii) the post-steps. Finally, the *StronglyGP* declares an input type to each node and organizes the primitives in three tiers corresponding to preprocessing, thresholding segmentation, and postprocessing. The results demonstrate that StronglyGP performs better than *TwoStageGP* and *CoevoGP.*

GP has also been used for Evolutionary Neural Architecture Search (ENAS), which aims to optimize the topology of convolutional neural nets. It has been applied to both image classification [14,17,20] and image segmentation tasks [6]. In this case, the function set includes related neural net operators, such as the creation of convolutional layers, dense layers, and non-linear pattern connections.

As aforementioned, a consequence of the bloat effect is the increment of the size and height of individuals. Therefore, several approaches have been explored to reduce the complexity of the models. Liang *et al.* [13] compare results employing a weighted sum of classification accuracy and a complexity measure and a multi-objective approach using the NSGA-II [4] and SPEA2 algorithms [21]

The majority of works that perform the segmentation do not use medical images. Furthermore, the set of functions in some works is robust. This leads to the creation of complex and large models, which require a long execution time.

3   Proposed Approach

3.1   GP - Image Feature Construction

The proposal uses the tree structure to construct and extract features from the grayscale CT medical images. The objective is to find the appropriate functions to construct and extract features. After that, the transformed images are segmented using two different methods: a fixed threshold value and an adaptive threshold (Otsu's method).

The overall algorithm is shown in Algorithm 1. Firstly, it receives a list of grayscale images and their corresponding ground-truth masks (G.T.), already segmented by an expert. Secondly, the initial population is generated with the Full method [11], where all individuals are created randomly with the same depth in all branches. In the third step, the fitness value is computed based on the segmentation performance and assigned to each individual of the initial population. In the loop, the tournament selection mechanism generates a list of parent individuals to create offspring through crossover and mutation operators. The sub-tree crossover and uniform mutation are employed as variation operators. The former exchanges the subtrees generated by a crosspoint on each parent individual. The latter replaces a subtree with another one randomly generated. The generated offspring are evaluated with the fitness function, and then the $\mu + \lambda$ strategy is adopted as a replacement mechanism. μ represents the number of individuals to select for the next generation during the evolutionary process, while λ represents the number of children created per generation. The process continues until it reaches a stop criteria, in this case, a maximum number of generations. Finally, it returns the modified images by operators found in the best individual of the evolved population.

Algorithm 1: GP-IFC

Input: Set of images (Grayscale and G.T.).
Output: Modified grayscale images and segmented images
Initial population
Eval population
while *Generations < Max Generations* **do**
  | Parent selection
  | Creation of new individuals using crossover and mutation operators
  | Eval new individuals
  | Survival selection
end

3.2   Function and Terminal Set

The function set includes two arithmetic functions, which are addition and subtraction. The four morphologic operators are opening, closing, erosion, and dilation. The structure element to perform the operations uses a kernel 7×7 in the shape of a disk [19]. Moreover, two filters are included: the Gaussian filter and the Canny filter. The parameters for those functions are set with default values [19]. It also contains a function that inverts the image's values, i.e., subtracting 255 minus the image's pixel values. It returns the absolute value of the operation. We proposed a function denoted as *mask* that operates as follows: As input, it receives the grayscale image. Then, a value between 0 and 255 was randomly selected as a threshold. According to the threshold value, all the values of the image matrix that are less or equal to the threshold take the value of 0. The unique terminal is the image, which is the input to the model to construct and extract meaningful features. Table 1 shows the functions and terminal set.

Table 1. Function set and terminals of GP

Function	Definition
Add	$(x + y)$
Substraction	$(x - y)$
Erosion	Morphologic operation
Dilation	Morphologic operation
Opening	Morphologic operation
Closing	Morphologic operation
Mask (proposed)	Pixel operation
Gaussian Filter	Filter
Canny	Filter
Invert Values	$\lvert 255 - I\ (x,y) \rvert$
Terminals	
Im	Grayscale image (2-D)

4   General Experiment Design

The dataset contains 100 grayscale images of CT scans of patients affected by COVID-19 and corresponding G.T with a variable resolution, ranging from 153×277 to 400×439 pixels. The dataset is split into 70% for training and 30% for testing.

The parameters calibrated experimentally were population and number of generations, detailed in the next section. The remaining are taken from the literature due to the efficiency. The crossover and mutation rates are set to 0.8 and 0.2, respectively [11]. The operator crossover is *one-point*, and mutation is the *uniform* [11]. The selection method is a tournament with a size of 5. The minimum and maximum depth of the initial population are set to 2 and 7, respectively, while the maximum depth of the generated population along generations is set to 17 [12]. μ is set to 100 and λ is set to 50.

The proposal is implemented in Python 3 and DEAP library [10]. Image processing functions are supported by OpenCV and Skimage. The experiments were run on a PC with Intel Core i7-8750H CPU @2.2 GHz and 16 GB RAM.

5   Results and Analysis

Two experiments have been conducted in this work. The first experiment is to calibrate the parameters of the population and the number of generations of the evolutionary process. When the number of generations and population size are set, two different based-threshold segmentation methods are compared.

5.1   Experiment 1

For the first experiment, the algorithm was run five times for 100, 150, and 200 individuals. Furthermore, it was run five times for the number of generations, with different values of 100, 150, and 200. Therefore, a total of thirty executions were conducted. The $F_1 - score$ metric, also known as Dice coefficient, is employed as a fitness function, expressed by:

$$F_1 - score = \frac{2 \times TP}{2 \times TP + FN + FP}$$

where TP is the total of true positive cases, FN is the total of false negatives and FP is the total of false positives. The metric was used as [12]. A value near 0 means the segmentation is poor; otherwise, a value near 1 means the segmentation is good. In this first experiment, the segmentation was performed using a global threshold with a fixed value. By simply selecting a threshold, such as the intermediate value in the grayscale range (128), it is possible to leave the discriminative power to the individuals of GP using the proposed function set.

The first parameter obtained was the population size. The comparison is supported by a Kruskal - Wallis test with a significance level of 95% to see if there is a difference between the results. The p-$value = 0.0804$ discarded a significant difference among the population size. Consequently, to reduce computational resources, the population size is 100 individuals. Secondly, the following parameter tuning was the number of generations. The algorithm was also run five times per number of generations. Using the same statistical test under the same significance level of 95%, the p-$value$= 0.4819. It determines that there is no significant difference between the three configurations. Therefore, the number of generations is set to 100 to maintain a reasonable execution time.

With the population size and maximum number of generations chosen, Table 2 shows the results of Experiment 1 of ten independent executions with the parameters already set. The column Run is the number of runs performed. The values in the columns *Threshold* and *Otsu* represent the average ± the standard deviation of the metric $F_1 - score$ over the test set. As we can see, the execution five achieves the highest value with 0.5912 of $F_1 - score$. The lowest value of ten runs is 0.3459. The average of ten runs is 0.4563 ± 0.07 Std. Dev.

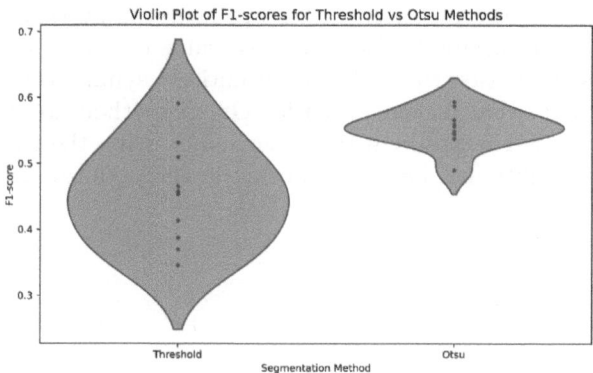

Fig. 1. Comparison between the Threshold and Otsu's segmentation method of ten executions.

Table 2. Results and statistics of ten executions using the threshold and Otsu segmentation methods.

Run	$F_1 - score$	
	Threshold	Otsu
1	0.45412 ± 0.25	0.56644 ± 0.122
2	0.37047 ± 0.16	0.55700 ± 0.120
3	0.46636 ± 0.11	0.54906 ± 0.229
4	0.53227 ± 0.10	0.54900 ± 0.120
5	$\mathbf{0.5912 \pm 0.13}$	0.56017 ± 0.233
6	0.45859 ± 0.25	0.58745 ± 0.223
7	0.38749 ± 0.11	0.49031 ± 0.105
8	0.34591 ± 0.11	0.53841 ± 0.158
9	0.41363 ± 0.13	$\mathbf{0.59394 \pm 0.123}$
10	0.51022 ± 0.13	0.54517 ± 0.115
Statistics		
Best	0.5912	0.59394
Worst	0.34591	0.4903
Average	0.45320	0.55369
Median	0.45635	0.55302
Std.Dev	0.076	0.028
$p-$**value**	0.0036	

Figure 1 shows a violin plot with the results of ten runs of both segmentation methods. The threshold method shows a dispersed distribution and a visible variability along the F1-score axis. The results go from 0.34 to the 0.59 F1-score value. The green violin is a little asymmetric, and the width region is near and lower than the average value of 0.45. The violin for the Otsu method shows a lower dispersion and the majority values are near the average value of 0.553. There exists a lower variability. However, the value of run 7 is not close to the median. That is the reason the violin is not entirely symmetric.

The results show that features used for Otsu's method are more stable and perform at a low distribution. On the other hand, using the threshold method creates more variability and high distribution in the performance.

The statistical test supports this.

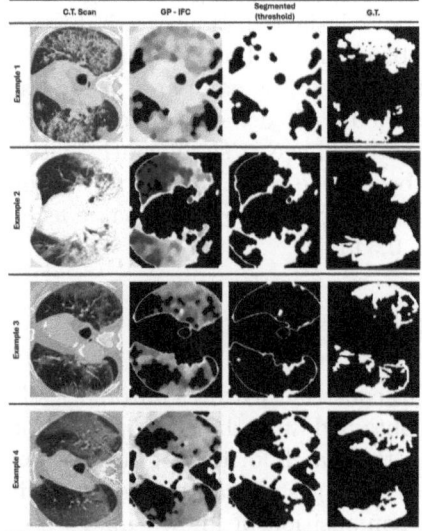 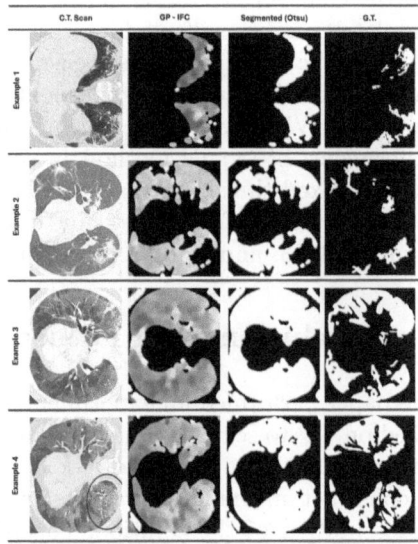

Fig. 2. Results of the feature extraction and the segmentation using the threshold at 128.

Fig. 3. Results of the feature extraction and the segmentation using the Otsu's method.

Figure 2 shows four examples of the results. The column *C.T. Scan* is the original image. Column *GP-IFC* is the feature constructed of applying the operators in the original image. *Segmented* corresponds to the segmented image of the GP-IFC features result using the threshold method. *C.T.* are the images segmented by hand, meaning they are the objective. As we can observe in the second and third rows, the result of the features constructed in the images can deal well with the background. However, when segmenting using the threshold, the image is sometimes inverted-converting the background to an interest region as in rows one and four. The main reason for this behavior is that the pixel values are greater than 128, and some images in the background contain close values as the region of interest.

5.2 Experiment 2

Once the number of generations and population size are set, the next step is to know which segmentation method can be used for the automatic feature construction. Otsu's method segments the images in this experiment instead of only using a fixed threshold value. The proposal was run ten times with ten different seeds. The $F_1 - score$ metric also is used as the fitness function.

The performance of Otsu's method to segment the images significantly increases the algorithm's performance. Quantitative and qualitative results

achieve better values regarding the $F_1 - score$ metric. Table 3 shows four qualitative results with the original image, the feature constructed, the binary image,
and the G.T. As we can see in three segmented images, the lungs are segmented
completely, and the affected interest region is not considered. However, in Experiment 1, the background is well-marked as black. According to the statistics
shown in Table 2, where experiment 1 vs experiment 2 is compared using the
the Mann-Whitney U test, the $p - value$ is 0.0036. It demonstrates that the
performance of Otsu's method in segmenting the image is significantly better
than the fixed threshold-based method. The best value was obtained in the run
nine, achieving a $F_1 - score$ value of 0.5939. The worst performance value using
Otsu is 0.4903, and using a fixed threshold is 0.345.

6   Conclusions and Future Work

The paper performed an experimental study to construct features automatically
based on GP to segment images. The results obtained after the experimentation
demonstrate the capability of GP to deal with the search for the set of functions
to construct meaningful features from the CT images. Because the syntactic
representation of the solutions consists of a composition of known image processing operators, interpretability and reproducibility are among the advantages
of genetic programming. In this way, a human can easily visually interpret and
simulate the final solution. The path for the features constructed and extracted
is known according to the segmentation method.

On the other hand, the experimentation allowed us to determine the parameters of population size and the number of generations for the evolutionary process. Taking this into account, these parameters found can be set for further
experimentation to increase the performance of the segmentation.

Furthermore, the results show that GP-IFC can automatically construct
important features from the grayscale images according to the segmentation
method. Additionally, they demonstrate that Otsu's method can improve the
proposed algorithm's performance. However, modifications are needed to deal
with those images that contain variations in the background, where the segmentation performs poorly.

The advantages of based-global threshold segmentation techniques are their
intuitiveness and straightforward implementation. Consequently, the distinguishing capability between classes is mainly left to the operators described by individuals. However, more complex techniques may lead to better segmentation
performance. Therefore, as future work, techniques that consider the local properties of the images, such as the local Otsu's method, can be further studied.
Additionally, more complex models, such as the ID3 or support vector machine,
can be considered, although it is crucial to weigh an increment in computational
cost. Finally, the further study of multi-objective techniques may reduce the
solution's complexity.

References

1. Ai, T., et al.: Correlation of chest ct and rt-pcr testing for coronavirus disease 2019 (covid-19) in China: a report of 1014 cases. Radiology **296**(2), E32–E40 (2020). https://doi.org/10.1148/radiol.2020200642, pMID: 32101510

2. Ain, Q.U., Xue, B., Al-Sahaf, H., Zhang, M.: Multi-tree genetic programming with a new fitness function for melanoma detection. In: 2019 IEEE Congress on Evolutionary Computation, CEC 2019 - Proceedings, pp. 880–887, June 2019. https://doi.org/10.1109/CEC.2019.8790282

3. Bi, Y., Xue, B., Zhang, M.: An evolutionary deep learning approach using genetic programming with convolution operators for image classification. In: 2019 IEEE Congress on Evolutionary Computation, CEC 2019 - Proceedings, pp. 3197–3204, June 2019. https://doi.org/10.1109/CEC.2019.8790151

4. Deb, K., Pratap, A., Agarwal, S., Meyarivan, T.: A fast and elitist multiobjective genetic algorithm: nsga-ii. IEEE Trans. Evolut. Comput. **6**, 182–197 (2002). https://doi.org/10.1109/4235.996017

5. Evans, B., Al-Sahaf, H., Xue, B., Zhang, M.: Evolutionary deep learning: A genetic programming approach to image classification. In: 2018 IEEE Congress on Evolutionary Computation, CEC 2018 - Proceedings, September 2018. https://doi.org/10.1109/CEC.2018.8477933

6. Fuentes-Tomás, J.A., Mezura-Montes, E., Acosta-Mesa, H.G., Márquez-Grajales, A.: Tree-based codification in neural architecture search for medical image segmentation. IEEE Trans. Evolut. Comput.**28**(3), 597–607 (2024). https://doi.org/10.1109/TEVC.2024.3353182

7. Herrera-Sánchez, D., Acosta-Mesa, H.G., Mezura-Montes, E.: Auto machine learning based on genetic programming for medical image classification. In: Calvo, H., Martínez-Villaseñor, L., Ponce, H., Zatarain Cabada, R., Montes Rivera, M., Mezura-Montes, E. (eds.) Advances in Computational Intelligence. MICAI 2023 International Workshops, pp. 349–359. Springer Nature Switzerland, Cham (2024). https://doi.org/10.1007/978-3-031-51940-6_26

8. Herrera-Sánchez, D., Acosta-Mesa, H.G., Mezura-Montes, E., Márquez-Grajales, A.: Shifting color space for image classification using genetic programming. In: Proceedings of the Genetic and Evolutionary Computation Conference Companion, pp. 283–286. GECCO '24 Companion, Association for Computing Machinery, New York, NY, USA (2024). https://doi.org/10.1145/3638530.3654430

9. Hiroyasu, T., Nunokawa, S., Yamaguchi, H., Koizumi, N., Okumura, N., Yokouchi, H.: Algorithms for automatic extraction of feature values of corneal endothelial cells using genetic programming. In: 6th International Conference on Soft Computing and Intelligent Systems, and 13th International Symposium on Advanced Intelligence Systems, SCIS/ISIS 2012, pp. 1388–1392 (2012). https://doi.org/10.1109/SCIS-ISIS.2012.6505152

10. Kim, J., Yoo, S.: Software review: deap (distributed evolutionary algorithm in python) library. Genet. Program. Evol. Mach. **20**, 139–142 (2019). https://doi.org/10.1007/S10710-018-9341-4/FIGURES/1, https://link.springer.com/article/10.1007/s10710-018-9341-4

11. Koza, J.R.: Human-competitive results produced by genetic programming. Genet. Program. Evol. Mach. **11**, 251–284 (2010). https://doi.org/10.1007/s10710-010-9112-3

12. Liang, J., Wen, J., Wang, Z., Wang, J.: Evolving semantic object segmentation methods automatically by genetic programming from images and image processing operators. Soft Computing **24**, 12887–12900 (2020). https://doi.org/10.

1007/S00500-020-04713-1, https://link.springer.com/article/10.1007/s00500-020-04713-1. 2020 24:17

13. Liang, Y., Zhang, M., Browne, W.N.: Multi-objective genetic programming for figure-ground image segmentation. In: Ray, T., Sarker, R., Li, X. (eds.) ACALCI 2016. LNCS (LNAI), vol. 9592, pp. 134–146. Springer, Cham (2016). https://doi.org/10.1007/978-3-319-28270-1_12

14. McGhie, A., Xue, B., Zhang, M.: GPCNN: evolving convolutional neural networks using genetic programming. In: 2020 IEEE Symposium Series on Computational Intelligence, SSCI 2020, pp. 2684–2691, December 2020. https://doi.org/10.1109/SSCI47803.2020.9308390

15. Pedrino, E.C., et al.: Automatic construction of image operators using a genetic programming approach. In: International Conference on Intelligent Systems Design and Applications, ISDA, pp. 636–641 (2011). https://doi.org/10.1109/ISDA.2011.6121727

16. Song, A., Ciesielski, V.: Fast texture segmentation using genetic programming. In: 2003 Congress on Evolutionary Computation, CEC 2003 - Proceedings, vol. 3, pp. 2126–2133 (2003). https://doi.org/10.1109/CEC.2003.1299935

17. Suganuma, M., Shirakawa, S., Nagao, T.: A genetic programming approach to designing convolutional neural network architectures. In: GECCO 2017 - Proceedings of the 2017 Genetic and Evolutionary Computation Conference, vol. 8, pp. 497–504, July 2017. https://doi.org/10.1145/3071178.3071229

18. Vojodi, H., Fakhari, A., Moghadam, A.M.E.: A new evaluation measure for color image segmentation based on genetic programming approach. Image Vis. Comput. **31**, 877–886 (2013). https://doi.org/10.1016/J.IMAVIS.2013.08.002

19. Walt, S.V.D., et al.: Scikit-image: image processing in python. PeerJ **2014**, e453 (2014). https://doi.org/10.7717/PEERJ.453/FIG-5, https://peerj.com/articles/453

20. Zhu, Y., et al.: GP-CNAS: convolutional neural network architecture search with genetic programming, November 2018. https://doi.org/10.48550/arxiv.1812.07611, https://arxiv.org/abs/1812.07611v1

21. Zitzler, E., Laumanns, M., Thiele, L.: Spea2: improving the strength pareto evolutionary algorithm (2001). https://doi.org/10.3929/ETHZ-A-004284029

Color Quantification in Common Bean Landraces Using a Supervised Learning Technique

José-Luis Morales-Reyes[1]($\boxtimes$) (iD), Elia-Nora Aquino-Bolaños[1] (iD),
and Héctor-Gabriel Acosta-Mesa[2] (iD)

[1] Centre for Food Research and Development, University of Veracruz, Veracruz, Mexico
jluismorey@hotmail.com
[2] Artificial Intelligence Research Institute, University of Veracruz, Veracruz, Mexico

Abstract. In common beans, color is a physical property related to nutritional compounds that are beneficial to health; in this work, we explored the quantification of color to know the different proportions of color in samples of common bean landraces. This research is significant as it sheds light on the importance of color quantification in nutritional analysis. A dataset was created with colorimetric information of seeds of seven different colorations that was used for the k-nearest neighbors algorithm for assigned labels; 168 images of bean landraces of heterogeneous color were processed to assign labels to each one; the labels generated for the seeds of each bean landrace allowed us to quantify the color; 121 samples the 100% of grain were correctly labeled, on the other hands, in 47 bean samples, the labels assigned to some seeds did not describe to the color of the seed. The results show that it is possible to quantify the color by classifying the set of representative seeds of bean landraces.

Keywords: Common bean landraces · Quantification · Classification · Color distribution · k-NN

1 Introduction

Common beans are a widely cultivated and consumed legume, based on genetic evidence in Mexico as the center of origin, diversification, and domestication [1]. There are many wild beans and varieties of common bean domesticated called common bean landraces, many of which are grown for local farmers for private consumption in rural communities [2, 3].

The common beans contain various nutritional components beneficial to health, such as proteins, fiber, polyphenols, and flavonoids, among other compounds related to antioxidant activity, which prevent cardiovascular effects or chronic degenerative diseases, as well as other conditions related to metabolic syndrome [4–8].

The bean landraces are adapted to different conditions, soil conditions, altitudes, and agricultural practices in each region. The genetic diversity is composed of varying color

L. Martínez-Villaseñor et al. (Eds.): MICAI 2024 Workshops, LNAI 15465, pp. 167–178, 2025.
https://doi.org/10.1007/978-3-031-83882-8_16

combinations related to various chemical concentrations [9]. The colorimetric characteristics of bean landraces are multiple; in the case of heterogeneous color landraces, that
is, the mixture of seeds of different colorations, if several samples of the same variety
of landraces are obtained, the samples will not present the same number of seeds of the
same coloration, it has been observed that the colorations are maintained in different
proportions [3].

The colorimetric properties of bean landraces are directly related to their chemical
composition; for this reason, the study of chemical compounds related to coloration
has been the research subject. The quantification of color will allow us to know the
percentages of the contribution of the different colorations of the bean seeds that conform
to the sample of bean landraces. To represent bean landraces, a set of seeds determined
by a certain weight is required for chemical analysis [10].

Several authors report different methods of characterization and classification, the
use of the color of each common bean, and the calculation of the values of the average,
or the use of the values of variance, skewness, and channel kurtosis; others authors
complemented with morphological information such as size and shape, the color spaces
used are RGB, HSV, and CIE L*a*b*. Supervised learning algorithms like k nearest
neighbor (k-NN), support vector machine (SVM), artificial neural network (ANN), and
Convolutional neural networks (CNN) are also used [11–15].

A previous related work reports the common beans classification by individual sample using a reduced color characterization [11–15]; the [16] reported the color characterization using joint probability distributions represented by one two-dimensional
histogram to characterize the color of a set of seeds; moreover, in [17] The color
characterization of the set of seeds is reported using three two-dimensional histograms.

To our knowledge, the quantification of color in common bean landraces has not yet
been explored; our contribution to the aims of the present work is:

1. To characterize the color of a seed common bean landraces, we propose to represent
   the color using three two-dimensional histograms created from the information in the
   CIE L*a*b* color space.
2. There is evidence that the k-NN algorithm can be used to classify using one two-
   dimensional histogram; this proposal will be trained using three two-dimensional
   histograms with the color information of seeds of bean landraces and used to identify
   each seed to assign labels in samples of bean landraces and to account for the total
   of colorations present in a sample.

The document's structure follows: Sect. 2 presents the material and methods for quantifying the color in bean landraces. Section 3 displays the results of the proposed methodology and describes the discussions based on those results. Finally, Sect. 4 provides the
conclusions and future work.

2   Materials and Methods

2.1   Color Quantification

A set of seeds represents a bean landrace sample, so they are calorimetrically variable. Digital image processing was performed to identify seed by seed and assigned labels to quantify the color proportions utilizing the set of labels obtained to determine the colorations present. The following series of steps were carried out for color quantification:

1. Image acquisition and image processing: First was the image acquisition of a set of seeds that represent each bean landrace; subsequently, it was converted image RAW to TIFF format necessary for the segmentation algorithm.
2. The segmentation result is a binary image necessary to locate the regions of interest. Label-connected components were required to find each seed's area to obtain information.
3. For each bean landrace, color characterization for each seed was three two-dimensional histograms.
4. The k-NN algorithm, known in data classification, was employed to predict the correct labeling of unobserved data, in this case, the seeds of each bean landraces. This step is crucial in the color quantification process. Previously, a base of knowledge for the algorithm was created, which is necessary for label prediction. The set of labels obtained corresponds to the set of seeds that make up a sample of bean landrace, and arithmetic operations are performed to calculate the percentages of the color proportions.
5. The meticulous procedure outlined above, crucial for quantifying the color proportions in a bean landrace sample, holds significant value in agricultural science and food. Figure 1 visually represents the steps involved, highlighting the impact of our research.

2.2   Bean Landraces

This work used 168 different common bean landraces of heterogeneous color. These landraces correspond to three domesticated species (*Phaseolus Vulgaris L, Phaseolus lunatus, and Phaseolus Coccineous*). Each landrace is conformed with seeds coats of different colorations (red, black, yellow, brown, pink, and purple), and each landrace was represented with a weight of 60 g. In Fig. 2, there are some samples of bean landraces.

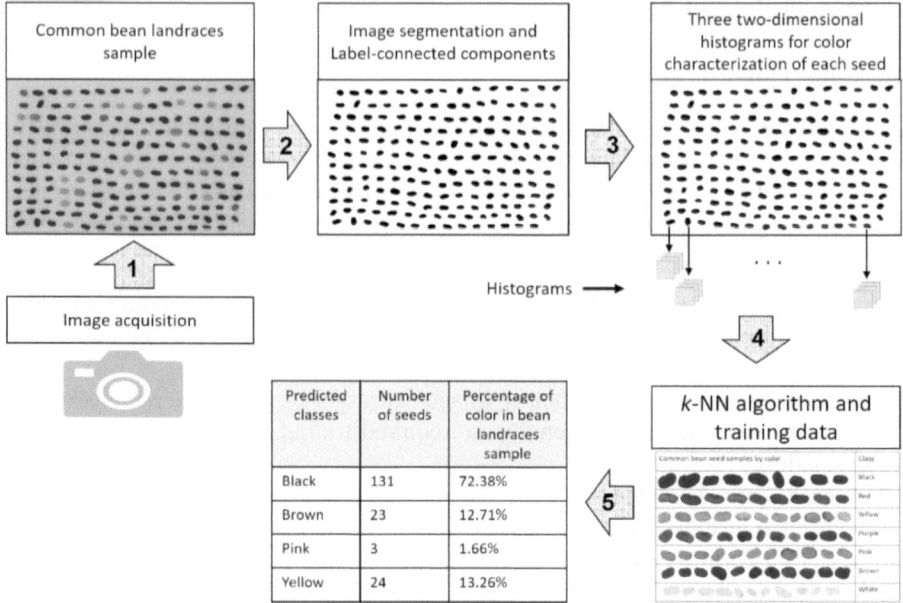

Fig. 1. Workflow for color quantification on common bean landraces.

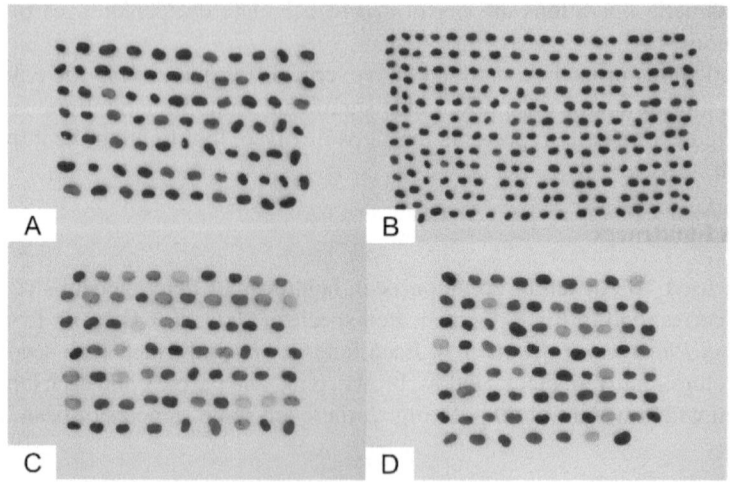

Fig. 2. For common bean landrace samples, from A to D, the set seed of each sample corresponds to 60 of a mixture of bean landrace seeds.

2.3 Acquisition System and Image Segmentation

The illumination setup has controlled lighting that maintains uniform illumination, mitigates glare, and reduces specular brightness for shiny bean landraces; it consists of an aluminum box with the camera installed in the upper section, and eight bulbs provide illumination inside. Additionally, the acquisition system has a color image reproduction workflow crucial for reproducing results using a computer vision system; the acquisition system used in this work is reported by [16].

The aluminum box is equipped with a sliding platform for placing the seed samples from each landrace, and blue was chosen to contrast the background color of the seed: the acquisition process involved using a digital camera, specifically the SONY ILSE 3500. The RAW images were acquired at a resolution of 5456 × 3632 pixels and, as a process part, were converted to the TIFF format with an sRGB color space using Darktable version 3.2 software. Consequently, the standard ICC profile was substituted with a custom ICC profile created using MATLAB software. The purpose of those above was to convert the sRGB to CIE L*a*b* color space.

The image processing techniques were applied to obtention regions of interest, and a color-based region-growing segmentation algorithm was employed [18]. The algorithms start with selecting a seed pixel; the region grows in the area where neighbor pixels share similitude determined to mean a criterium. Equation 1 was used to measure the similitude among pixels.

$$\Delta E_{ab} = \sqrt{\left(L_p - L_s\right)^2 + \left(a_p - a_s\right)^2 + \left(b_p - b_s\right)^2} \tag{1}$$

where: ΔE_{ab} is a CIE color accuracy computed value L*a*b*, (L, a, b) of seed pixel (s) and (L, a, b) of neighboring pixel (p).

2.4 Color Characterization Using a Probability Mass Function

The color characterization used to represent the apiece seed of each bean landrace was the probability mass function, and the information on regions of interest was obtained from the CIE L*a*b* color space. Therefore, given two discrete random variables, X and Y, , a joint probability distribution or probability mass function (PMF) is defined as $f(X, Y) = P(X = x, Y = y)$, where $f(X, Y)$ represents the occurrence probability of x and y values, subject to the following conditions.

1. $f(x, y) \geq 0$ *for all* (x, y),
2. $\sum_x \sum_y f(x, y) = 1$
3. $P(X = x, Y = y) = f(x, y)$

Based on the information presented above, it is possible to generate histograms utilizing the chromaticity channel data from the CIE L*a*b*. The 8-bit images contain 256 shades of gray, making it possible to create a histogram with 256 bins of pictures of 5456 × 3632 pixels resolution. To account for the dominance of the chromaticity channels a* and b* [−128, 127] in CIE L*a*b*, the shift was computed by summing the values of each pixel along with the absolute value of the lower limit. L* channel represents the perceptual lightness, defined as 0 and 100 for black and white color, respectively. The values in the L* channel were normalized for rescaling to [0–255].

The color characterization used to characterize each seed of bean landraces was constructed using three two-dimensional histograms: histogram of a* and b*, histogram of L* and a*, and histogram of L* and b* obtained from the CIE L*a*b* color space.

2.5   Color Reference

As a part of learning, disposing of a dataset with information on the diverse colorations that present the common bean landraces is essential. As a first approach, this work created a dataset with the color information of seeds selected from the different bean landraces. In this way, for the generation of the color palette with the support of the expert, a set of seeds was selected from the different images; digital image processing was applied to locate the seeds and obtain the color information. To dispose of the distinct shade variations, seeds of similar coloration were selected for each color group (see Fig. 3). Each seed's color characterization was done using three two-dimensional histograms.

Common bean seed samples by color	Class
	Black
	Red
	Yellow
	Purple
	Pink
	Brown
	White

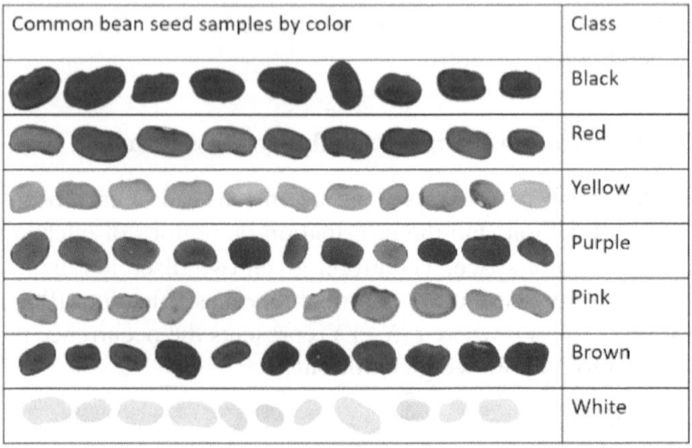

Fig. 3.   The seed for color reference represents training data for the machine learning algorithm.

2.6   Classification Algorithm

The k-NN is a supervised learning algorithm; its operation is based on the location of points in the multidimensional space. Classifying a new observation takes the class label among the k-nearest neighbors; first, it must calculate the observation distance with all the points that belong to the training set. The city block distance is appropriate for comparing histograms (see Eq. 2); the k-nearest neighbors are selected, and the highest frequency class will be assigned to the unknown observation through voting [19].

$$d_T(a, b) = \sum_{j=1}^{n} |a_j - b_j| \tag{2}$$

where: $a = (a_1, a_2, \ldots, a_n)$, $b = (b_1, b_2, \ldots, b_n)$, a and b are histograms, and n-points represent the value frequencies.

The k-NN algorithm stores a training dataset instead of going through a training phase. Therefore, it is suitable for storing the color reference dataset. So also, the calculation for classification or prediction occurs when an unknown observation arrives, and using a metric, the distance between the query point and the base points is measured to find its neighbors and assign a label.

In this work, the function fitcknn of MATLAB software 2023b represents a k-NN classification model; nine neighbors were considered for a more significant number of neighbors, and the parameter Distance-Weight with the value squared inverse was assigned.

2.7 Experiment Designed

The procedure described in the methodology was performed for color quantification in this work; previously, the information on the different colorations of each bean landrace was obtained, representing the knowledge for the k-NN algorithm. The experiment involves processing each bean landrace's image to get each seed's color information and to know the labels assigned to them by the k-NN algorithm; finally, the set of labels was grouped to get the amounts that share the same labels.

3   Results and Discussion

The results will be expressed numerically in two groups: the first corresponds to bean landraces correctly color quantified, and the second corresponds to bean landraces not correctly quantified.

Table 1. Quantities of correctly quantified and incorrectly quantified bean landraces.

Correctly quantified	Incorrectly quantified
121	47

Examples of correctly quantified bean landraces are shown in Table 2.
Examples of incorrectly quantified bean landraces are shown in Table 3.

Table 2. Some samples of bean landraces were correctly quantified. The legend below in the image is the name of the bean landrace.

Images of common bean landraces	Labels	Number of seeds	Percentage of color in bean land-races sample
 PC-004-TOO-003-R1-C2	Black Brown Purple	60 6 11	77.92% 7.79% 14.28%
 PC-005-TOO-004-R1-C1	Black	61	100%
 PC-005-TOO-124-R3-C2	Black Purple	59 11	84.28% 15.71%
 PC-007B-TOO-007-R1-C1	Black Brown Purple Red Yellow	45 1 53 8 46	29.41% 0.65% 34.64% 5.22% 30.06%
 PC-013-TOO-013-R1-C1	Black Brown Pink Purple Red Yellow	8 8 10 43 4 5	10.25% 10.25% 12.82% 55.12% 5.12% 6.41%

Table 3. Samples of bean landraces were erroneously quantified. The legend below in the image is the name of the bean landrace.

Images of common bean landraces	Labels	Number of seeds	Percentage of color
PC-008-TOO-008-R1-C1	Black Brown Purple Red White Yellow	117 1 21 5 1 9	75.97% 0.64% 13.63% 3.24% 0.64% 5.84%
PC-011-TOO-011-R1-C2	Black Brown Pink Purple Red White Yellow	4 28 7 16 1 1 12	5.79% 40.58% 10.14% 23.18% 1.44% 1.44% 17.39%
PC-022-TOO-275-R4-C1	Black Brown Pink	12 64 6	14.63% 78.04% 7.31%
PL-001-ZAA-166-R3-C1	Black Brown Pink Purple Red Yellow	5 9 1 7 2 152	2.84% 5.11% 0.56% 3.9% 1.13% 86.36%
PL-001-ZAA-228-R4-C1	Black Brown Purple Red	1 122 38 4	0.60% 73.93% 23.03% 2.42%

In all bean landrace samples, the seeds did not have assigned color identification labels due to the time it may take for the expert to identify them and the subjectivity this represents. To quantify the color automatically, it was necessary to elaborate on the color reference and knowledge base for the k-NN algorithm. The results of color quantification, as shown in Table 1, demonstrate the effectiveness of this reference. All the seeds of 121 samples of bean landraces were correctly identified, and the label assigned corresponded to the color of the seed; therefore, it was possible to fully quantify the colors of the bean landraces, while there were some errors in the prediction of color in a few bean landraces.

On the other hand, Table 2 shows five examples of bean landraces with correct color quantification; in each, the labels assigned to each seed correspond to its color, and the expert performed verification to check the proper assignment. The label column contains the different color labels that are obtained by grouping the labels assigned by the algorithm of classification, the number of seeds column includes the total of seeds that share the same label, and the percentage of the color column shows the percentages that each color group represents in the bean sample. The number of seeds calculated in each color group was divided by the total number of seeds in the sample, which should represent 100% of the total number of seeds in the sample.

In Table 3, we observe five examples of bean landraces that presented prediction errors, such as the prediction of white seeds in the landraces PC-008-TOO-008-R1-C1 and PC-011-TOO-011-R1-C2; some seeds assigned as yellow color are very similar to white seeds so that a prediction error may occur. Notably, these errors were minimal, less than 5% in the set of seeds of each local bean landrace.

In the case of the landraces with variegated seeds, there were more significant prediction errors. Variegated seeds represent a challenge because they contain pigments of different coloration, so quantifying color requires exploring other techniques.

This methodology can be used to assign color labels to bean landraces of various colorations, both homogeneous and heterogeneous. However, Using this method allows for reduced subjectivity and error reduction in the work of [16], which was reported as a color palette as a visual reference for assigning a superclass label; this work provides the basis for assigning labels efficiently.

4   Conclusions and Future Work

Color quantification is not just a process but a crucial step in understanding the nutritional compounds present in a sample of bean landraces. The proportions of color are directly related to these compounds, which are beneficial to health. Therefore, knowing these proportions provides a valuable reference for the compounds with a more significant presence, underscoring the importance of work. The color reference is a crucial element that enables the learning algorithm to accurately recognize the colorations in a bean landrace.

Color is a property of great importance since it is related to the chemical composition of bean landraces; in this work, we explored the quantification of the color samples, which will allow us to know the contribution of color and the presence of different chemical concentrations related to color.

Digital image processing techniques are essential to obtain the information from a set of seeds representative of bean landraces. In addition, using three two-dimensional histograms to characterize color allowed us to identify the different seeds of each landrace to quantify color in terms of percentage.

Our study demonstrates the potential of the k-NN algorithm in automating color identification. In future work, we propose three aspects: i) to further explore more techniques of machine learning to quantify color in bean landraces, particularly in the case of variegated seeds; ii) This approach can be used to quantify color in samples of other crops, and iii) this approach also opens up possibilities for to be used in assigning color-based class labels in different crops for classification tasks.

Acknowledgments. The first author acknowledges the National Council of Humanities, Sciences, and Technologies (CONAHCyT) of Mexico for granting support for the realization of this investigation through scholarship 712056 awarded for postdoctoral studies at the Centre for Food Research and Development at the University of Veracruz.

References

1. Gepts, P., Debouck, D.: Origin, domestication, and evolution of the common bean (Phaseolus vulgaris L.). (1991)
2. Vargas-Torres, A., Osorio-Díaz, P., Tovar, J., Paredes-López, O., Ruales, J., Bello-Pérez, L.A.: Chemical composition, starch bioavailability and indigestible fraction of common beans (Phaseolus Vulgaris L.). Starch Stärke **56**, 74–78 (2004). https://doi.org/10.1002/star. 200300205
3. Chávez-Servia, J.L., et al.: Diversity of common bean (Phaseolus vulgaris L.) Landraces and the nutritional value of their grains. In: Goyal, A.K. (ed.) Grain Legumes. InTech (2016). https://doi.org/10.5772/63439
4. Chen, J., Xu, B., Sun, J., Jiang, X., Bai, W.: Anthocyanin supplement as a dietary strategy in cancer prevention and management: a comprehensive review. Crit. Rev. Food Sci. Nutr. **62**, 7242–7254 (2022). https://doi.org/10.1080/10408398.2021.1913092
5. Guo, H., Ling, W.: The update of anthocyanins on obesity and type 2 diabetes: experimental evidence and clinical perspectives. Rev. Endocr. Metab. Disord. **16**, 1–13 (2015). https://doi. org/10.1007/s11154-014-9302-z
6. Tsuda, T.: Regulation of adipocyte function by anthocyanins; possibility of preventing the metabolic syndrome. J. Agric. Food Chem. **56**, 642–646 (2008). https://doi.org/10.1021/jf0 73113b
7. Wallace, T.C.: Anthocyanins in cardiovascular disease. Adv. Nutr. **2**, 1–7 (2011). https://doi. org/10.3945/an.110.000042
8. Zafra-Stone, S., Yasmin, T., Bagchi, M., Chatterjee, A., Vinson, J.A., Bagchi, D.: Berry anthocyanins as novel antioxidants in human health and disease prevention. Mol. Nutr. Food Res. **51**, 675–683 (2007). https://doi.org/10.1002/mnfr.200700002
9. Hernández-Delgado, S., et al.: Advances in genetic diversity analysis of phaseolus in Mexico. In: Caliskan, M., Oz, G.C., Kavakli, I.H., Ozcan, B. (eds.) Molecular Approaches to Genetic Diversity. InTech (2015). https://doi.org/10.5772/60029
10. Aquino-Bolaños, E., GarcaDaz, Y., ChavezServia, J., CarrilloRodrguez, J., VeraGuzman, A., HerediaGarcia, E.: Anthocyanins, polyphenols, flavonoids and antioxidant activity in common bean (Phaseolus vulgaris L.) landraces. Emir. J. Food Agric. **28**, 581 (2016). https://doi.org/ 10.9755/ejfa.2016-02-147

11. Kılıç, K., Boyacı, İH., Köksel, H., Küsmenoğlu, İ: A classification system for beans using computer vision system and artificial neural networks. J. Food Eng. **78**, 897–904 (2007). https://doi.org/10.1016/j.jfoodeng.2005.11.030
12. Araújo, S.A.D., Pessota, J.H., Kim, H.Y.: Beans quality inspection using correlation-based granulometry. Eng. Appl. Artif. Intell. **40**, 84–94 (2015). https://doi.org/10.1016/j.engappai.2015.01.004
13. Bianco, M.L., Grillo, O., Cremonini, R., Sarigu, M., Venora, V.: Characterisation of Italian bean landraces ("Phaseolus vulgaris" L.) using seed image analysis and texture descriptors. Aust. J. Crop Sci. **9**, 1022–1034 (2015)
14. Nasirahmadi, A., Behroozi-Khazaei, N.: Identification of bean varieties according to color features using artificial neural network. Span. J. Agric. Res. **11**, 670–677 (2013). https://doi.org/10.5424/sjar/2013113-3942
15. Venora, G., Grillo, O., Ravalli, C., Cremonini, R.: Identification of Italian landraces of bean (Phaseolus vulgaris L.) using an image analysis system. Scientia Horticulturae **121**, 410–418 (2009). https://doi.org/10.1016/j.scienta.2009.03.014
16. Reyes, J.L.M., Mesa, H.G.A., Bolanos, E.N.A., Meza, S.H., Ramirez, N.C., Servia, J.L.C.: Classification of bean (Phaseolus vulgaris L.) landraces with heterogeneous seed color using a probabilistic representation. In: 2021 IEEE International Autumn Meeting on Power, Electronics and Computing (ROPEC), pp. 1–7. IEEE, Ixtapa, Mexico (2021). https://doi.org/10.1109/ROPEC53248.2021.9668106
17. Morales-Reyes, J.-L., Aquino-Bolaños, E.-N., Acosta-Mesa, H.-G., Márquez-Grajales, A.: Estimation of anthocyanins in homogeneous bean landraces using neuroevolution. In: Calvo, H., Martínez-Villaseñor, L., Ponce, H., Zatarain Cabada, R., Montes Rivera, M., Mezura-Montes, E. (eds.) Advances in Computational Intelligence. MICAI 2023 International Workshops, pp. 373–384. Springer Nature Switzerland, Cham (2024). https://doi.org/10.1007/978-3-031-51940-6_28
18. Woods, R.E., Gonzalez, R.C.: Digital image processing third edition (2021)
19. de Batista, G.E.A.P.A., Silva, D.F.: How k-nearest neighbor parameters affect its performance. Anales JAIIO (2009)

Explainable AI Through Decision Trees for Black-Box Models Used to Support Bacterial Vaginosis Diagnosis

Rafael Rivera-López[1]([✉]), Juana Canul-Reich[2], Erick De la Cruz Hernández[3],
Héctor Gibrán Ceballos-Cancino[4], Efrén Mezura-Montes[5],
and Marco Antonio Cruz-Chávez[6]

[1] DSC, TecNM, Instituto Tecnológico de Veracruz, Veracruz, Mexico
rafael.rl@veracruz.tecnm.mx
[2] DACyTI, Universidad Juárez Autónoma de Tabasco,
Cunduacán, Mexico
juana.canul@ujat.mx
[3] DAMC, Universidad Juárez Autónoma de Tabasco,
Comalcalco, Mexico
erick.delacruz@ujat.mx
[4] IFE, Tecnológico de Monterrey, Monterrey, Mexico
ceballos@tec.mx
[5] IIIA, Universidad Veracruzana, Xalapa, Mexico
emezura@uv.mx
[6] CIICAP, Universidad Autónoma del Estado de Morelos, Cuernavaca, Mexico
mcruz@uaem.mx

Abstract. Bacterial vaginosis is a significant public health concern affecting the reproductive health of sexually active women, underscoring the need for tools that enhance diagnostic accuracy. This study introduces a model-agnostic framework that provides explainability for black-box models used to diagnose bacterial vaginosis. The framework combines the predictive power of opaque models, such as artificial neural networks and random forests, with the transparency and interpretability of decision trees. By leveraging both the training data and the predictions from black-box models, the framework constructs decision trees without requiring access to the internal mechanics of the black-box models. Experimental results demonstrate that this approach generates decision trees more accurately than those created from the original training dataset.

Keywords: Explainable Artificial Intelligence · Black-box models · Decision trees

1 Introduction

The impact of machine learning across various fields of human activity is undeniable. Its use in decision-support tools for sectors like economics [5], healthcare

L. Martínez-Villaseñor et al. (Eds.): MICAI 2024 Workshops, LNAI 15465, pp. 179–189, 2025.
https://doi.org/10.1007/978-3-031-83882-8_17

[20], and science [10] has expanded our understanding of complex phenomena and processes previously beyond our reach. In particular, the application of machine learning in medicine has been transformative, with significant advancements in areas such as information management [22], patient monitoring [34], disease diagnosis [36], and medical follow-up [32]. This manuscript focuses specifically on using machine learning to aid in diagnosing bacterial vaginosis, a condition considered a public health issue due to its potential to cause long-term reproductive health complications if not detected early [16]. However, integrating machine learning in medicine requires accurate and interpretable models. While black-box models, such as neural networks, offer impressive predictive capabilities, they often lack transparency, making their use in medical decision-making more challenging [27]. Explainable Artificial Intelligence has emerged as a promising solution, providing insight into how these models arrive at their predictions [35].

This study introduces an explainable machine learning approach in which the outcomes of two black-box models (one artificial neural network and a random forest) are used to train a decision tree. The goal is to improve the understanding of clinical study results related to diagnosing bacterial vaginosis. Experimental findings suggest that this approach enhances diagnostic accuracy, surpassing the performance of decision trees trained solely on the original dataset.

The rest of this manuscript is organized as follows: Sect. 2 introduces the Bacterial Vaginosis conditions and their impact on public health. Section 3 describes the importance of providing explainability to black-box models in critical areas such as medicine. The related studies using machine learning methods for Bacterial Vaginosis diagnosis are summarized in Sect. 4. The proposed scheme in this study is clarified in Sect. 5, and the experimental study is detailed in Sect. 6. Finally, Sect. 7 is related to conclusions and future work.

2   Bacterial Vaginosis Diagnosis

Bacterial Vaginosis (BV) is a common vaginal condition resulting from an imbalance in the vaginal microbiota, involving a decrease in *lactobacilli* and an overgrowth of other bacteria such as *Gardnerella vaginalis*. It is not a sexually transmitted infection but is associated with sexual activity [29]. Although the presence of symptoms such as a thin, white, or gray vaginal discharge, a strong fishy odor (especially after sex), vaginal itching, and burning during urination is indicative of this condition, some women with BV may not show any symptoms [7]. Although Amsel Criteria with Gram Stain is the most common diagnostic method, alternative tests such as Nucleic Acid Amplification tests, pH test strips, and Point-of-Care tests can report more accurate BV diagnosis [1]. The accurate diagnosis and timely treatment of BV are crucial because untreated BV can increase the risk of acquiring sexually transmitted infections (STIs), including Human Immunodeficiency Virus (HIV), and can lead to obstetric complications, such as preterm birth. Furthermore, BV is prone to recurrence, so follow-up might be necessary [23].

BV is a global public health problem. In 2019, the World Health Organization (WHO) estimated the occurrence of 357 million new cases per year, mainly in developing countries where studies on their prevalence are limited [8]. In Mexico, some studies have investigated BV diagnosis in Tabasco [31], Puebla [17], and Veracruz [33] states and Mexico City [15].

3   Explainable Artificial Intelligence

Explainable Artificial Intelligence (XAI) aims to develop techniques or methods that, when used with black-box models, enable us to interpret how data is utilized and explain their predictions [19]. In this context, explainability refers to the ability to provide meaning to predictions, while interpretability refers to the capability to understand how the model processes data to produce results (also known as comprehensibility) [18]. There are several ways to achieve the explainability of black-box models, including inspecting their internal logic, providing justifications for their results, and creating white-box models that replicate their behavior [13]. XAI methods can be categorized as either model-specific or model-agnostic. The last case is related to those methods that can be applied to any black-box model. Furthermore, these methods can provide four explanation types: (1) extracting logic rules from the black-box model (decision trees [14] or rule sets [24]), (2) making sense of black-box internal elements [3], (3) assigning credit to input features [21], and (4) providing similar examples to the black-model predictions [2]. Finally, XAI methods can be globally explainable when they understand the complete logic of a model or locally explainable when they provide meaning to a specific prediction [37].

4   Related Methods

Bech and Foster [4] conducted one of the first studies where machine learning techniques were applied to BV characteristics data. They use two rRNA gene datasets with healthy and BV-diagnosed women. The diagnosis is based on Amsel's criteria and Nugent's score. They apply three machine-learning techniques: genetic programming (GP), Random Forest (RF), and Logistic Regression (LR). Furthermore, they deconstruct the classification models to identify important features of the microbial community. The same two-class datasets have been used in other studies: Celeste et al. [6] analyze the ethnic disparity in diagnosing asymptomatic BV using LR, RF, Support Vector Machine (SVM), and Multi-layer Perceptron (MLP) classifiers. Pérez-Gomez et al. [25] apply a decision tree (DT) and the Relief algorithm to select the top fifteen features to build a reduced dataset that is classified using SVM and LR algorithms.

On the other hand, other BV datasets have been used in the literature. Rodriguez et al. [28] apply one clustering algorithm to identify four risk behavior groups with a dataset of 135 BV-diagnosed women using the Nugent criteria. Furthermore, one multi-class dataset with 201 Mexican non-pregnant and sexually active women is used by De la Cruz-Ruiz et al. [9] to extract a set of association rules that describe the bacteria interacting to develop BV. The dataset includes women diagnosed as BV positive, BV negative, and those whose BV can not be identified (undetermined). This dataset is analyzed by Salvador-González et al. [30] to select a near-optimal association rules subset to identify the possible pathogen combinations coexisting in BV-diagnosed women.

Finally, Drew et al. [11] induce a DT from a dataset with 300 women subjected to a BV molecular test. The BV diagnosis is based on the Gram stain results.

5 Proposed Method

The study presents a model-agnostic framework for XAI that explains black-box models used in BV diagnosis. The proposed approach utilizes predictions from a black-box model. Firstly, the opaque model is created using a training set. Next, the training set is evaluated using the black-box model, and the original class labels are replaced with the model's predictions. Finally, a DT is induced using the re-labeled training set.

In this study, we use two different black-box models: an RF and one MLP. The DT is induced using a recursive partitioning approach based on the C4.5 algorithm. C4.5 uses the gain-ratio as partition criteria and the trained DT is pruned using the error-based pruning approach [26].

6 Experimental Study

The data used for this study correspond to medical information from 106 pregnant women who attended the Metabolic and Infectious Diseases Research Laboratory of the Universidad Juárez Autónoma de Tabasco, México and gave their written consent[1].

The dataset has seven numerical attributes identifying both lactobacilli bacteria that regulate the vaginal microbiota and anaerobic bacteria mainly related to BV infection: The values stores are the density values of *Lactobacillus crispatus*, *Lactobacillus jensenii*, and *Lactobacillus iners*, as well as the level of *BV associated bacterium 2*, *Megasphaera type 1*, *Atopobium vaginae*, and *Gardnerella vaginali*. The data are labeled into 15 cases with BV positive, 65 with normal microbiota (BV negative), and 26 with an indeterminate diagnosis. The classification of the patients is defined by the procedure proposed by Sánchez et al. [31].

In this study, two experiments are conducted to compare the performance of the model-agnostic framework. The first experiment uses an MLP as a black-box model, and the second utilizes an RF as the opaque model. The DTs are induced using the J48 method, a Java implementation of the C4.5 method provided by the Weka library [12]. The hyperparameter values used for the MLP are the

---

[1] The data collection procedure was conducted according to international standards for responsible publication of (COPE) and registered (protocol No. UJAT-20160006) and approved by the Institutional Review Board of the Universidad Juarez Autonoma de Tabasco [31].

default ones in Weka, except for the number of training iterations, which is set to 1000. The RF consists of one hundred base models, each with a maximum depth of five levels. All DTs are pruned using the Error-Based Pruning approach. Finally, the threshold value to determine whether a node should be labeled as a leaf node is set to two instances. These experiments are implemented in Java.

These experiments utilize a repeated stratified 10-fold cross-validation (CV) approach. In a 10-fold CV, the dataset is randomly divided into ten approximately equal, non-overlapping folds. For each iteration, the model is trained on nine folds and evaluated on the remaining one. This process is repeated until every fold has been used as the test set exactly once. This method ensures that the model is evaluated on all instances of the dataset, providing a reliable estimate of its performance. The 10-fold CV is repeated ten times, and the average test accuracy is used as the final performance metric for the model.

6.1   DT from ANN

Table 1 shows the results of using an MLP as a black-box model. The MLP predictions using the training set are utilized to create a new DT (Agnostic-J48 in this manuscript). The table compares the training and test accuracies of each model. The first column indicates the number of experiment repetitions, while the following two columns (labeled as 2 and 3 in Table 1) show the MLP results. Columns 4 and 5 depict the training and test accuracies of the DT induced from the original dataset using the J48 algorithm. The last two columns (6 and 7) show the results for the training and test sets after relabeling the dataset. In the last case, the class label of the instances used in the training phase is replaced by the prediction from the black-box model.

Table 1. Average accuracies obtained in the first experiment.

Number of experiment repetition[1]	MLP		J48		Agnostic J48	
	Training[2]	Test[3]	Training[4]	Test[5]	Training[6]	Test[7]
1	95.39	74.53	83.23	77.36	93.19	80.19
2	96.02	78.30	84.80	76.42	92.24	78.30
3	95.07	82.08	83.96	75.47	93.08	83.02
4	96.33	76.42	86.58	75.47	93.08	79.25
5	95.70	74.53	85.01	76.42	92.87	78.30
6	95.70	76.42	83.75	77.36	93.29	80.19
7	95.70	76.42	84.17	73.58	91.82	78.30
8	95.81	77.36	85.32	75.47	92.87	81.13
9	95.91	77.36	86.27	71.70	93.82	76.42
10	95.91	77.36	83.86	71.70	92.87	76.42
Average Accuracy	**95.75**	**77.08**	**84.70**	**75.09**	**92.91**	**79.10**

In Table 1, it can be observed that the MLP model has the best training accuracy. The DT induced with the original dataset has a lower predictive power, although its test accuracy is only 2% lower than that of the MLP. Finally, it can also be observed that the DT generated with the relabeled dataset has the best test accuracy, even 2% higher than the results of the black box model. This behavior indicates that altering the class label with the MLP predictions improves the DT's training power (with the relabeled data), allowing it to make better test predictions than the DT generated with the original dataset.

6.2   DT from RF

Table 2 shows the results of using an RF as a black-box model. This table is organized similarly to Table 1, but in this case, columns 2 and 3 show the RF results.

Table 2. Average accuracies obtained in the second experiment.

Number of experiment repetition[1]	RF		J48		Agnostic J48	
	Training[2]	Test[3]	Training[4]	Test[5]	Training[6]	Test[7]
1	95.70	78.30	84.17	75.47	92.87	79.25
2	96.02	78.30	84.49	73.58	92.98	78.30
3	96.33	78.30	85.01	74.53	92.77	79.25
4	96.54	80.19	85.12	76.42	92.24	81.13
5	96.44	78.30	84.91	71.70	92.14	78.30
6	96.54	80.19	87.21	72.64	93.19	76.42
7	96.02	81.13	83.96	77.36	92.45	83.96
8	95.70	78.30	82.81	76.42	92.77	80.19
9	96.44	80.19	85.43	75.47	93.08	80.19
10	95.49	78.30	85.43	73.58	92.35	79.25
Average Accuracy	**96.12**	**79.15**	**84.85**	**74.72**	**92.68**	**79.62**

The results in this table show a similar behavior to those in Table 1. It can be observed that as the black box model has better results, the agnostic model also improves. In this case, it is also observed that the DT generated with the relabeled dataset has the best test accuracy.

6.3   Discussion

According to the experimental results described previously, it is observed that the DT generated with the original dataset has a lower predictive power than those generated by the black-box models, a behavior that is well documented in the literature. The recursive partitioning strategy followed by a pruning procedure can reduce the DT predictive power. On the other hand, re-labeling the

training data with the black-box model predictions improves both the DT training and test predictions. Here, we can intuit that altering the class label has
the effect of improving the training power, which increases test accuracy. This
behavior is because the training data is regrouped based on its internal relationships, not the initially provided class label. It can also be observed that the
predictive power of the black-box model affects the white-box model. The RF
results are better than the MLP results, affecting the DT results created with
the re-labeled data.

The previous analysis only considers the accuracy of the produced models,
but this strategy aims to provide explainability to black box models. Figures 1,
2, and 3 show an example of the DTs induced with this strategy. Figure 1 shows
a DT generated with the complete original dataset using J48. It can be observed
in this figure that the model is compact but has several incorrectly labeled cases.

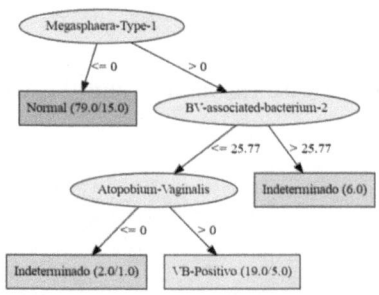

Fig. 1. DT induced the original dataset.

On the other hand, Fig. 2 shows the DT generated with the re-labeled dataset
with the MLP predictions. This DT is more complex but with greater predictive
power than that of Fig. 1. This DT has a similar behavior to the MLP used to
re-label its data, so it can be used to explain the results produced by the MLP
model.

Finally, Fig. 3 represents the DT induced with the data re-labeled by the RF
predictions. In this case, the DT has two additional internal nodes, producing
better predictive power. It is important to note that these models have been
generated with the hyperparameters set as default in the library used, yet an
improvement in the predictive power of the agnostic model is observed.

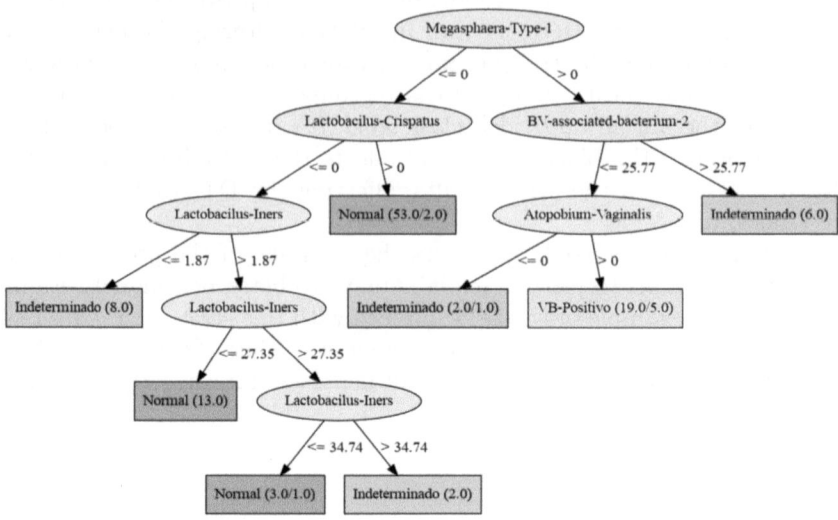

Fig. 2. DT induced with the dataset and the MLP predictions.

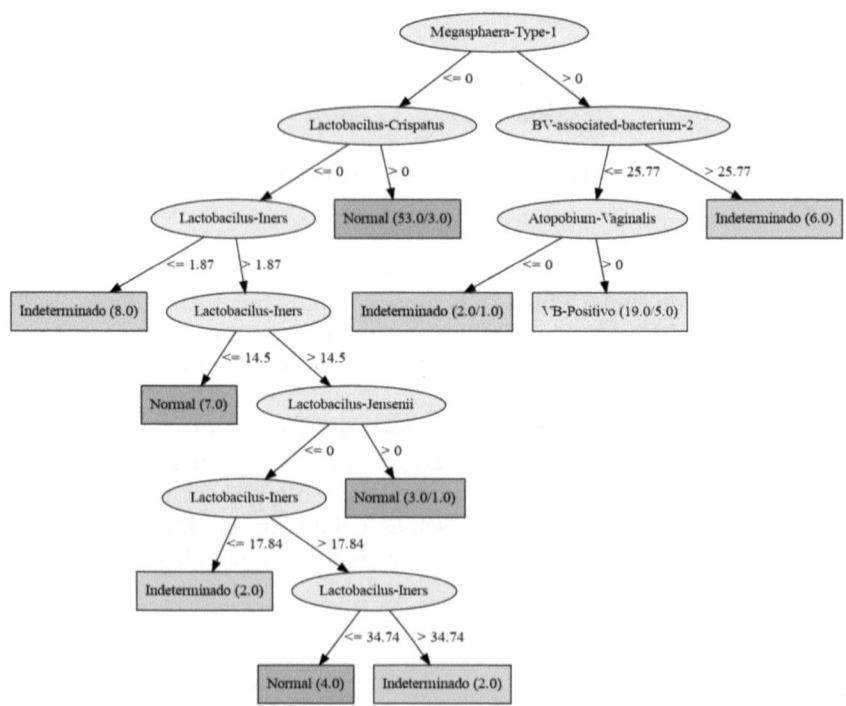

Fig. 3. DT induced with the dataset and the RF predictions.

7   Conclusions and Future Work

This manuscript introduces a model-agnostic framework for extracting DTs from black-box models' predictions. This approach utilizes the predictions of the black-box model on the training data to re-label the dataset, which is then used to train a white-box model.

The experimental results demonstrate that this framework produces more accurate DTs than those induced with the original training set. These improved models can serve as viable options for replicating the behavior of black-box models. Consequently, this enhances the applicability of black-box models in the medical field by providing a mechanism for high explainability in data management, which can aid in determining better medical diagnoses.

In the context of the data utilized in this study, this framework enables specialists treating female reproductive system disorders to make more accurate diagnoses. This improvement is expected to redu6ce the VB and, more significantly, patient health complications.

Future work will involve further experimentation with different configurations of black-box models and analyzing cases with indeterminate diagnoses.

References

1. Abou Chacra, L., Fenollar, F., Diop, K.: Bacterial vaginosis: what do we currently know? Front. Cell. Infect. Microbiol. **11** (2022). https://doi.org/10.3389/fcimb. 2021.672429
2. Atakishiyev, S., Salameh, M., Yao, H., Goebel, R.: Explainable artificial intelligence for autonomous driving: a comprehensive overview and field guide for future research directions. IEEE Access (2024). https://doi.org/10.1109/ACCESS.2024. 3431437
3. Bassi, P.R., Dertkigil, S.S., Cavalli, A.: Improving deep neural network generalization and robustness to background bias via layer-wise relevance propagation optimization. Nat. Commun. **15**(1), 291 (2024). https://doi.org/10.1038/s41467-023-44371-z
4. Beck, D., Foster, J.A.: Machine learning techniques accurately classify microbial communities by bacterial vaginosis characteristics. PLOS ONE **9**(2), 1–8 (2014). https://doi.org/10.1371/journal.pone.0087830
5. Çağlayan Akay, E., Yılmaz Soydan, N.T., Kocarık Gacar, B.: Bibliometric analysis of the published literature on machine learning in economics and econometrics. Soc. Netw. Anal. Min. **12**(1), 109 (2022). https://doi.org/10.1007/s13278-022-00916-6
6. Celeste, C., et al.: Ethnic disparity in diagnosing asymptomatic bacterial vaginosis using machine learning. NPJ Digit. Med. **6**(1), 211 (2023). https://doi.org/10. 1038/s41746-023-00953-1
7. Chen, X., Lu, Y., Chen, T., Li, R.: The female vaginal microbiome in health and bacterial vaginosis. Front. Cell. Infect. Microbiol. **11** (2021). https://doi.org/10. 3389/fcimb.2021.631972
8. Medina-De la Cruz, O., et al.: Vaginal infections: a public health problem in Mexico. Revista Médica de la Universidad Autónoma de Sinaloa REVMEDUAS **14**(1), 71–86 (2024)

9. De la Cruz-Ruiz, F., Canul-Reich, J., De la Cruz-Hernández, E., Rivera-Lopez, R.: Analysis of bacterial association patterns that trigger bacterial vaginosis. Int. J. Comb. Optim. Probl. Inform. **13**(4), 83–102 (2022). https://www.ijcopi.org/ojs/article/view/301

10. Deiana, A.M., et al.: Applications and techniques for fast machine learning in science. Front. Big Data **5**, 787421 (2022). https://doi.org/10.3389/fdata.2022.787421

11. Drew, R.J., Murphy, T., Broderick, D., O'Gorman, J., Eogan, M.: An interpretation algorithm for molecular diagnosis of bacterial vaginosis in a maternity hospital using machine learning: proof-of-concept study. Diagn. Microbiol. Infect. Dis. **96**(2), 114950 (2020). https://doi.org/10.1016/j.diagmicrobio.2019.114950

12. Frank, E., Hall, M., Witten, I.: The WEKA Workbench. Online Appendix (2016). https://www.cs.waikato.ac.nz/ml/weka/Witten_et_al_2016_appendix.pdf

13. Guidotti, R., Monreale, A., Ruggieri, S., Turini, F., Giannotti, F., Pedreschi, D.: A survey of methods for explaining black box models. ACM Comput. Surv. (CSUR) **51**(5), 1–42 (2018). https://doi.org/10.1145/3236009

14. Hada, S.S., Carreira-Perpiñán, M.Á., Zharmagambetov, A.: Sparse oblique decision trees: a tool to understand and manipulate neural net features. Data Min. Knowl. Discov. 1–40 (2023). https://doi.org/10.1007/s10618-022-00892-7

15. Hernández-Rodríguez, C., Romero-González, R., Albani-Campanario, M., Figueroa-Damián, R., Meraz-Cruz, N., Hernández-Guerrero, C.: Vaginal microbiota of healthy pregnant Mexican women is constituted by four lactobacillus species and several vaginosis-associated bacteria. Infect. Dis. Obstet. Gynecol. **2011**(1), 851485 (2011). https://doi.org/10.1155/2011/851485

16. Jayaram, P.M., Mohan, M.K., Konje, J.: Bacterial vaginosis in pregnancy-a storm in the cup of tea. Eur. J. Obstet. Gynecol. Reprod. Biol. **253**, 220–224 (2020). https://doi.org/10.1016/j.ejogrb.2020.08.009

17. Jiménez-Flores, G., Flores-Tlalpa, J., Ruiz-Tagle, A.C., Villagrán-Padilla, C.L.: Evaluación de los métodos utilizados para el diagnóstico de vaginosis bacteriana en el Hospital Regional ISSSTE Puebla. CienciaUAT **14**(2), 62–71 (2020)

18. Kamath, U., Liu, J.: Explainable artificial intelligence: an introduction to interpretable machine learning. Springer (2021). https://doi.org/10.1007/978-3-030-83356-5

19. Kumar, D., Mehta, M.A.: An overview of explainable AI methods, forms and frameworks. In: Mehta, M., et al. (eds.) Explainable AI: Foundations, Methodologies and Applications, pp. 43–59. Springer, Cham (2023). https://doi.org/10.1007/978-3-031-12807-3_3

20. Kumari, J., Kumar, E., Kumar, D.: A structured analysis to study the role of machine learning and deep learning in the healthcare sector with big data analytics. Arch. Comput. Methods Eng. **30**(6), 3673–3701 (2023). https://doi.org/10.1007/s11831-023-09915-y

21. Mastropietro, A., Feldmann, C., Bajorath, J.: Calculation of exact Shapley values for explaining support vector machine models using the radial basis function kernel. Sci. Rep. **13**(1), 19561 (2023). https://doi.org/10.1038/s41598-023-46930-2

22. Mirzaei, A., Aslani, P., Schneider, C.R.: Healthcare data integration using machine learning: a case study evaluation with health information-seeking behavior databases. Res. Soc. Adm. Pharm. **18**(12), 4144–4149 (2022). https://doi.org/10.1016/j.sapharm.2022.08.001

23. Morris, M., Nicoll, A., Simms, I., Wilson, J., Catchpole, M.: Bacterial vaginosis: a public health review. Br. J. Obstet. Gynaecol. **108**(5), 439–450 (2001). https://doi.org/10.1016/S0306-5456(00)00124-8

24. Obregon, J., Jung, J.Y.: RuleCOSI+: rule extraction for interpreting classification tree ensembles. Information Fusion **89**, 355–381 (2023). https://doi.org/10.1016/j.inffus.2022.08.021

25. Pérez-Gómez, J.F., Canul-Reich, J., Hernández-Torruco, J., Hernández-Ocaña, B.: Predictor selection for bacterial vaginosis diagnosis using decision tree and relief algorithms. Appl. Sci. **10**(9) (2020). https://doi.org/10.3390/app10093291

26. Quinlan, J.R.: C4.5: Programs for Machine Learning. Morgan Kaufmann, USA (1993)

27. Quinn, T.P., Jacobs, S., Senadeera, M., Le, V., Coghlan, S.: The three ghosts of medical AI: can the black-box present deliver? Artif. Intell. Med. **124**, 102158 (2022). https://doi.org/10.1016/j.artmed.2021.102158

28. Rodriguez, V.J., et al.: Using unsupervised machine learning to classify behavioral risk markers of bacterial vaginosis. Arch. Gynecol. Obstet. **309**(3), 1053–1063 (2024). https://doi.org/10.1007/s00404-023-07360-7

29. Russo, R., Karadja, E., Seta, F.D.: Evidence-based mixture containing Lactobacillus strains and lactoferrin to prevent recurrent bacterial vaginosis: a double blind, placebo controlled, randomised clinical trial. Beneficial Microbes **10**(1), 19–26 (2019). https://doi.org/10.3920/BM2018.0075

30. Salvador-González, M.C., Canul-Reich, J., Rivera-López, R., Mezura-Montes, E., De la Cruz-Hernandez, E.: Evolutionary selection of a set of association rules considering biological constraints describing the prevalent elements in bacterial vaginosis. Math. Comput. Appl. **28**(3) (2023). https://doi.org/10.3390/mca28030075

31. Sanchez-Garcia, E.K., Contreras-Paredes, A., Martinez-Abundis, E., Garcia-Chan, D., Lizano, M., De la Cruz-Hernandez, E.: Molecular epidemiology of bacterial vaginosis and its association with genital micro-organisms in asymptomatic women. J. Med. Microbiol. **68**(9), 1373–1382 (2019). https://doi.org/10.1099/jmm.0.001044

32. Sánchez-Puente, A., et al.: Machine learning to optimize the echocardiographic follow-up of aortic stenosis. JACC: Cardiovasc. Imaging **16**(6), 733–744 (2023). https://doi.org/10.1016/j.jcmg.2022.12.008

33. Vázquez Torres, T.P.: Vaginosis bacteriana en amenaza de parto pre término en el Hospital de Alta Especialidad de Veracruz. Ph.D. thesis, Universidad Veracruzana. Facultad de Medicina. Región Veracruz (2018)

34. Vimal, S., Vadivel, M., Baskar, V.V., Sivakumar, V., Srinivasan, C.: Integrating IoT and machine learning for real-time patient health monitoring with sensor networks. In: 2023 4th International Conference on Smart Electronics and Communication (ICOSEC), pp. 574–578. IEEE (2023). https://doi.org/10.1109/ICOSEC58147.2023.10275890

35. Whig, P., Kouser, S., Bhatia, A.B., Nadikattu, R.R., Sharma, P.: Explainable Machine Learning in Healthcare, pp. 77–98. Springer, Cham (2023). https://doi.org/10.1007/978-3-031-38036-5_5

36. Wu, Y., Wu, M.: Biomedical data mining and machine learning for disease diagnosis and health informatics. Bioengineering **11**(4) (2024). https://doi.org/10.3390/bioengineering11040364

37. Zhang, Y., Tiňo, P., Leonardis, A., Tang, K.: A survey on neural network interpretability. IEEE Trans. Emerg. Top. Comput. Intell. **5**(5), 726–742 (2021). https://doi.org/10.1109/TETCI.2021.3100641

Improving Lactation Curve Estimation in Sheep: A Comparative Analysis of Machine Learning Algorithms Across Milk Recording Schemes

L. Guevara[1], F. A. Castro-Espinoza[2($\boxtimes$)], A. M. Fernandes[1], I. Nacarati-da-Silva[1], T. Oliveira[1], E. G. Salgado Hernández[3], and J. C. Angeles-Hernadez[3($\boxtimes$)]

[1] Centro de Ciências e Tecnologias Agropecuárias, Universidade Estadual do Norte Fluminense, Campos dos Goytacazes, Brasil
[2] Instituto de Ciencias Básicas e Ingeniería, Universidad Autónoma del Estado de Hidalgo, Hidalgo, México
fcastro@uaeh.edu.mx
[3] Departamento de Medicina y Zootecnia de Rumiantes, Facultad de Medicina Veterinaria y Zootecnia, Universidad Nacional Autónoma de México, Mexico City, Mexico
juanangeles@fmvz.unam.mx

Abstract. Estimation of lactation curve characteristics is essential for effective dairy sheep management, influencing nutrition, health and genetic improvement strategies. Traditional statistical methods often lack the capability to capture the complexity of milk production patterns, requiring innovative data analysis techniques. The aim of the current study is to evaluate the effectiveness of different Machine Learning (ML) algorithms - SMOreg, Linear Regression, M5 and Random Forest - in estimating key lactation curve parameters: Total Milk Yield (TMY), Peak Yield (PY) and Time to Peak Yield (TPY) under three different milk recording schemes. A total of 2,280 weekly records were used from a commercial sheep flock in Querétaro, Mexico. The results showed that the ML algorithms provided accurate estimates of lactation curve characteristics. We also found that the estimations of lactation curve characteristics were similar between the milk recording scheme using 20 records weekly (ST) and five records monthly (TF). In conclusion, the ML algorithms used can improve lactation curve estimation in dairy sheep and provide a cost-effective alternative for sheep production management by maximizing data utility while minimizing operational costs. Future research should focus on refining these algorithms for different sheep breeds and environmental conditions to further integrate them into practical farm management systems.

Keywords: Machine Learning · Lactation Curve · Sheep

1 Introduction

Estimation of lactation curve characteristics is essential for the effective management of dairy sheep flocks as it directly influences nutritional, health and genetic improvement strategies (Silvestre et al., 2006). The lactation curve represents the relationship

L. Martínez-Villaseñor et al. (Eds.): MICAI 2024 Workshops, LNAI 15465, pp. 190–199, 2025.
https://doi.org/10.1007/978-3-031-83882-8_18

between milk production and time and provides critical insight into the productivity and efficiency of dairy sheep. Accurate modelling of lactation curve is essential to optimize herd performance and promotes sheep production systems economically viable and environmentally responsible (Guevara et al., 2023).

Traditional statistical methods for estimating lactation curves are often based on mathematical models (e.g. Wood model), which have often shown deficiencies in describing atypical milk production patterns, especially in sheep (Val-Arreola et al., 2004). In addition, these traditional approaches have several other limitations, such as sensitivity to missing data, the need for normal distribution data, and over-parameterization when many predictors are used, which limits their use and fitting performance (Angeles-Hernandez et al., 2022). Dallago et al. (2019) compare a traditional approach based on multiple linear regression (MLR) with machine learning algorithms to estimate milk production in dairy cows. They found that Artificial Neural Networks (ANN) had the best performance. In addition, ANN were more flexible than MLR because it does not make assumptions about the data distribution, such as homoscedasticity and normality of the residual error.

As a result, there is an increasing demand for innovative data analysis techniques that can improve the accuracy of lactation curve estimation in the dairy industry. In this context, Machine Learning (ML) algorithms have emerged as powerful tools capable of modelling complex data sets and atypical milk production patterns. Computational models using novel artificial intelligence (AI) and ML algorithms offer objective, data-driven analyses that can be applied to various domains, including prediction, classification, image analysis, regression, object recognition, and anomaly detection (Alwadi et al., 2024). In dairy sheep, the performance of ML algorithms was compared with the Wood model, which is the most commonly used method for fitting lactation curves. This study shows that ML algorithms significantly improve the predictive ability for estimating total milk yield (TMY), peak yield (PY) and Time to Peak Yield (TPY) compared to the Wood model (Guevara et al., 2023).

ML algorithms have demonstrated success in predicting and modelling various biological processes, making them a promising approach for lactation curve analysis. ML techniques have the potential to improve traditional mathematical models by better capturing the variation between individual animals and the effects of environmental factors on milk production (Ince and Sofu, 2013). Therefore, the aim of this study is to evaluate the effectiveness of different ML algorithms in estimating key lactation curve parameters - Total Milk Yield (TMY), Peak Yield (PY) and Time to Peak Yield (TPY) - using data from three different sheep milk recording schemes with different availability of data. By improving the accuracy and efficiency of lactation curve estimation maximizing the used of available data, this research aims to contribute to improved management practices in dairy sheep production.

2   Materials and Methods

2.1   Database

A total of 2,280 weekly records corresponding to 156 lactations from sheep in a commercial herd located in Querétaro, Mexico, were used. The actual total milk yield (TMY) was estimated from the weekly milk production records using the centered day method (Sargent, 1968). Peak yield (PY) and time to peak yield (TPY) were identified by visual inspection of the lactation curves.

2.2   Milk Production Recording Schemes

Three new databases were generated from the original dataset. Each database corresponds to a milk production recording scheme, which was used to generate the input attributes for the ML algorithms. The input attributes for each scheme were: 1) first day of milk production recording (FDR), 2) lactation duration (LD), and 3) milk yield at the nth production record (MY1, MY2…MY20). The schemes were as follows:

- *SF scheme*: Attributes 1, 2, and the first five weekly milk production records (MY1, MY2…MY5).
- *ST scheme*: Attributes 1, 2, and the first twenty weekly milk production records (MY1, MY2…MY20).
- *TF scheme*: Attributes 1, 2, and the first five monthly milk production records (MY1, MY2…MY5).

2.3   Model Formulation

The model formulation consisted of mapping the input attributes for each scheme (SF: 1–7, ST: 1–22, TF: 1–7) to produce the output attributes (a-c), which corresponded to the characteristics of the lactation curve: a) TMY, b) PY, and c) TPY.

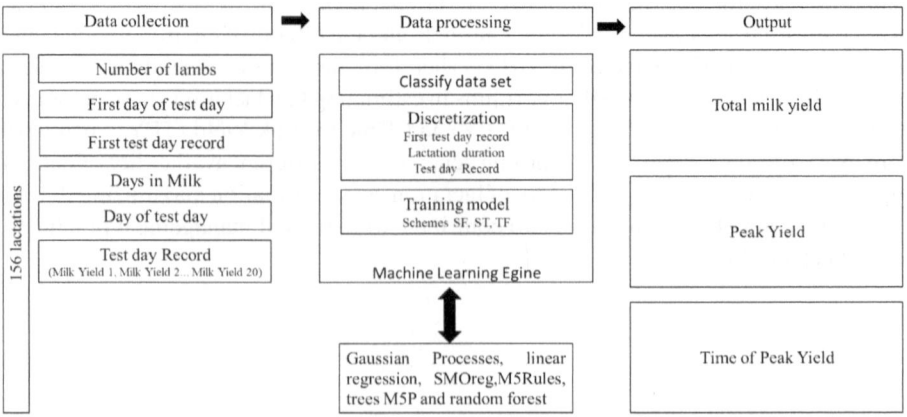

Fig. 1. Architecture of the model.

The model development involved training and testing models using ML algorithms for each output variable (a, b, c). The algorithms for model development were executed in the Waikato Environment for Knowledge Analysis software (WEKA, version 3.8.6). Subsequently, the six best algorithms were selected from the 28 available in WEKA: "functions.GaussianProcesses" (GP), "functions.LinearRegression" (LR), "functions.SMOreg" (SMO), "rules.M5Rules" (M5), "trees.M5P" (M5P), and "trees.RandomForest" (RF) (Fig. 1).

2.4   Evaluation of Goodness of Fit

The goodness of fit of the models was assessed using the following metrics:

1. *Coefficient of Correlation (r)*: this metric measures the strength and direction of the linear relationship between the predicted and the actual values.
2. *Mean Absolute Error (MAE)*: it represents the average absolute difference between the predicted values and the actual values. It measures the average magnitude of the errors without considering their direction. It's represented by:

$$MAE = (1/n)* \sum |yi - xi| \tag{1}$$

where yi and xi are the observed and estimated values respectively, and n is the total number of observations.
3. *Root Mean Square Error (RMSE)*: It is calculated by taking the square root for the average of the squared differences between the predicted values and the actual values. The RMSE provides a measure of the overall magnitude of the errors, giving more weight to larger errors.

$$RMSE = \sqrt{\left((1/n)* \sum (yi - xi)^2\right)} \tag{2}$$

where yi and xi are the observed and estimated values respectively, and n is the total number of observations.
4. *Relative Absolute Error (RAE)*: this metric is the ratio of the MAE to the mean of the actual values. It represents the average absolute difference between the predicted values and the actual values relative to the scale of the actual values.

$$RAE = (\Sigma|yi - xi|)/(\Sigma|\bar{y} - xi|) \tag{3}$$

where yi and xi are the observed and estimated values respectively, and $\bar{y}$ is the mean of the observed values.
5. *Relative Root Mean Square Error (RRSE)*: Similar to RAE, RRSE is the ratio of the RMSE to the mean of the actual values. It represents the average magnitude of the errors relative to the scale of the actual values.

$$RRSE = \sqrt{\left(\left(\sum (yi - xi)^2\right)/\left(\sum (\bar{y} - xi)^2\right)\right)} \tag{4}$$

where yi and xi are the observed and estimated values respectively, and $\bar{y}$ is the mean of the actual values.

The comparisons between the values of MY, PY and TPY estimated by the ML algorithms and the actual values were carried out by a variance analysis evaluating the effect of the algorithms, the milk recording scheme and their interaction as a fixed effect using the lm function in R. The post hoc comparisons were carried out using the *contrast* function in the "*eemeans*" package in the statistical language R (Lenth 2024). Plots showing the least square means and significance test were drawn to show the interaction between the ML algorithms and the milk recording schemes on the lactation curve characteristics.

3 Results

The results of selecting the best method for recording lactation data to more accurately estimate sheep lactation characteristics using ML algorithms are presented in Table 1. The ML algorithms LR, SMO, M5, and M5P, used to estimate milk production, show small differences in goodness-of-fit criteria across all methods (SF, RMSE = 0.15; ST, RMSE = 1.9; TF, RMSE = 0.77). For peak yield, all algorithms performed similarly in terms of fit, with the GP algorithm having the lowest performance (RMSE = 0.15–0.18 vs. 0.19–0.21). In contrast, for time to peak yield, the GP algorithm showed the best fit across all milk production recording methods, with minimal differences (SF, RMSE = 0.67; ST, RMSE = 1.71; TF, RMSE = 1.04) compared to the LR, SMO, M5, and M5P algorithms.

According to the goodness-of-fit criteria used, the ST and TF methods showed the best fit for total milk yield using the SMOreg algorithm. For peak yield, there were small differences in RMSE (0.0126) between the evaluated methods for the M5Rules algorithm, with the SF method achieving the lowest RMSE (0.151) and a high correlation (r = 0.87). This same method (SF) also showed the best goodness of fit (RMSE = 16.30) for the GP algorithm; however, the differences with the other methods were also considered minimal (1.09).

Regarding the characteristics of the lactation curve, the only scheme that showed differences between the algorithms and the actual values was the SF scheme (Fig. 1), where the best estimation of TMY was shown by the SMO algorithm (actual: 111.0 l vs. GP: 103.9 l). Contrary to what was observed in the other algorithms, the TMY was significantly underestimated (P = 0.003). In the ST and TF schemes, there are no significant differences between the actual TMY and that estimated by the ML algorithms (P > 0.05) (Table 2).

For PY, no significant differences between actual and estimated values were observed for the SF scheme (P = 0.05). However, in the ST scheme, most algorithms significantly underestimated the actual PY values (P = 0.001), with only the GP algorithm providing an accurate estimate for this characteristic of the lactation curve (actual: 0.97 l vs. GP: 1.0 l). In contrast, the GP algorithm showed a significant overestimation of close to four liters of PY when using the TF scheme database (Fig. 2). TMY was adequately estimated by most of the ML algorithms across the three milk recording systems (Fig. 3). However, the SMO algorithm was the only one that showed significant differences from the actual values (P < 0.001), consistently underestimating this characteristic of the lactation curve in all three schemes (Table 2).

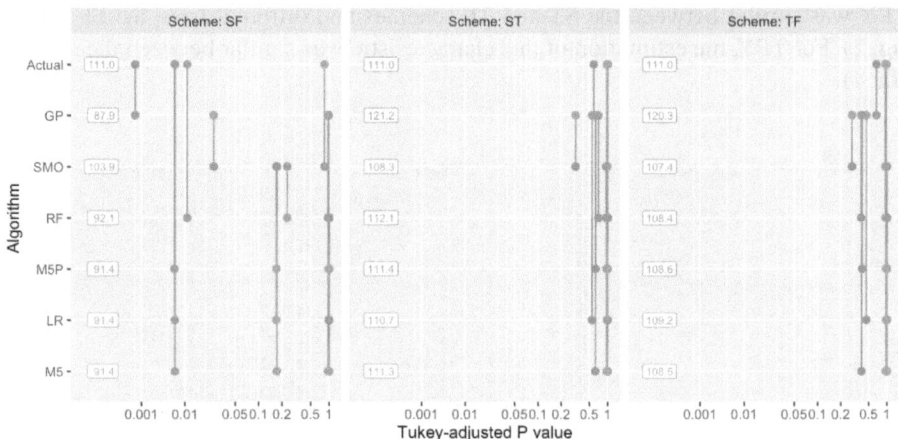

Fig. 2. Least squares means and pos-hoc comparison of Total Milk Yield of the interaction between the milk recording scheme and the Machine Learning algorithms. SF scheme (5 milk records weekly); ST scheme (20 milk records weekly); TF scheme (5 milk records monthly). GP, Gaussian processes; LR, Linear regression; SMOreg, SMOregression; M5, M5Rules; M5P, M5P tree; RF, Random Forest.

Assessing the effect of milk recording schemes on the estimation of lactation curve characteristics, our results revealed that the estimations of TMY were similar between them (Fig. 1). The same pattern was observed for PY characteristics, where the estimation

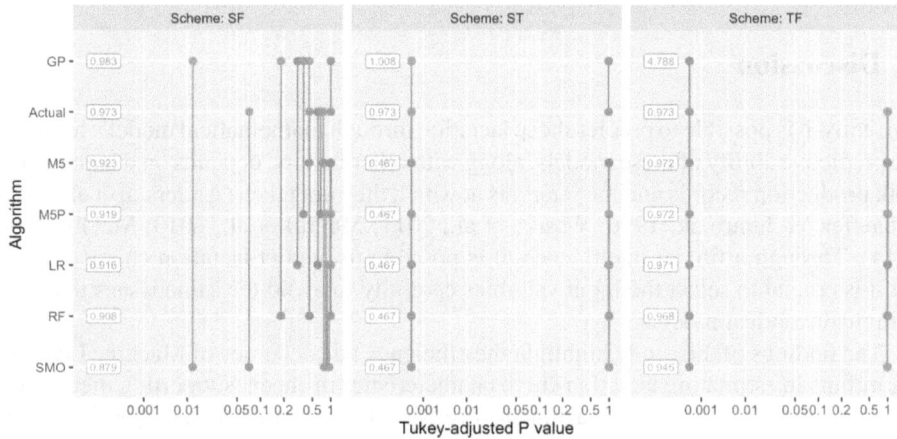

Fig. 3. Least squares means and pos-hoc comparison of Peak Yield of the interaction between the milk recording scheme and the Machine Learning algorithms. SF scheme (5 milk records weekly); ST scheme (20 milk records weekly); TF scheme (5 milk records monthly). GP, Gaussian processes; LR, Linear regression; SMOreg, SMOregression; M5, M5Rules; M5P, M5P tree; RF, Random Forest.

of PY was similar between the ST and TF schemes and different from the SF scheme (Fig. 2). For TPY, the estimation of this characteristic was similar between the schemes (Fig. 4).

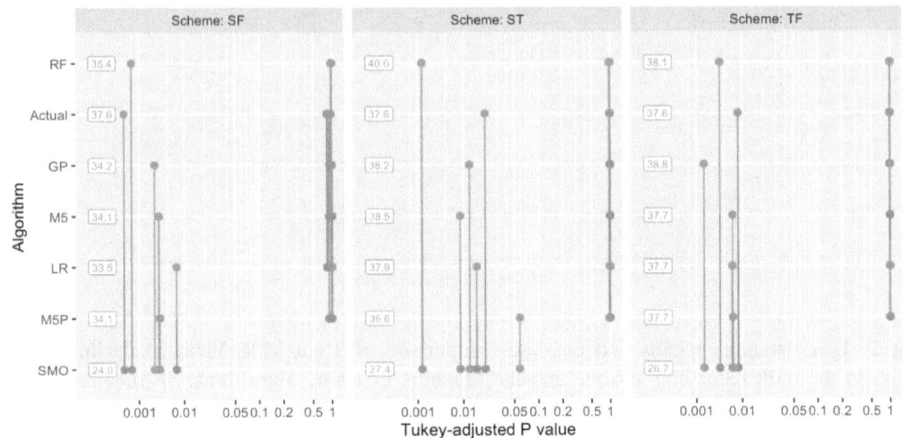

Fig. 4. Least squares means and pos-hoc comparison of Time to Peak Yield of the interaction between the milk recording scheme and the Machine Learning algorithms. SF scheme (5 milk records weekly); ST scheme (20 milk records weekly); TF scheme (5 milk records monthly). GP, Gaussian processes; LR, Linear regression; SMOreg, SMOregression; M5, M5Rules; M5P, M5P tree; RF, Random Forest.

4 Discussion

Currently, it is possible to predict sheep lactation through mathematical models; however, the predictive ability of these models, along with other factors, depends on the number of milk production records and the intervals at which they are taken (Anderson et al., 1989; Schaeffer °& Jamrozik, 1996; Wasike et al., 2011; McGill et al., 2013; McGill et al., 2014). Through artificial intelligence, it is possible to predict lactation characteristics, but it is crucial to select the input variables carefully to avoid the same issues that arise with mathematical models.

The findings of this study highlight the efficiency and accuracy of Machine Learning algorithms in estimating lactation curve characteristics in sheep, supporting their potential application in livestock management. The superior performance of the SMOreg algorithm across all data collection schemes, particularly in the ST and TF protocols, emphasizes its robustness in handling complex datasets, even in cases where noise or missing data may be present, as previously noted by Nguyen et al. (2020) and Mammadova & Keskin (2013).

The importance of selecting the appropriate data input variables was evident, as the prediction accuracy improved significantly with more comprehensive datasets, such as

Table 1. Goodness of fit of Machine Learning algorithms for lactation characteristics: Total Milk Yield, Peak Yield, and Time to Peak Yield in dairy sheep.

	Total milk yield						Peak yield						Time to peak yield					
	GP	LR	SMO	M5	M5P	RF	GP	LR	SMO	M5	M5P	RF	GP	LR	SMO	M5	M5P	RF
7 days, 5 test day records (SF)																		
r	0.79	0.94	0.94	0.94	0.94	0.91	0.72	0.83	0.83	0.87	0.86	0.86	0.70	0.70	0.70	0.70	0.70	0.38
MAE	24.66	11.88	11.87	12.02	12.02	14.68	0.17	0.13	0.13	0.11	0.11	0.11	11.39	11.18	10.02	11.13	11.13	16.56
RMSE	29.85	15.64	15.79	15.78	15.78	19.56	0.21	0.17	0.17	0.15	0.15	0.16	16.30	16.41	16.96	16.39	16.39	26.22
RAE (%)	61.90	29.83	29.80	30.16	30.16	36.85	67.66	50.29	49.41	43.61	44.22	42.06	59.97	58.85	52.76	58.58	58.58	87.18
RRSE (%)	63.61	33.34	33.65	33.63	33.63	41.69	70.33	55.45	56.38	49.93	50.51	51.50	70.68	71.15	73.57	71.09	71.09	113.71
7 days, 20 test day records (ST)																		
r	0.70	0.96	0.97	0.95	0.95	0.90	0.68	0.76	0.80	0.80	0.82	0.85	0.68	0.62	0.69	0.62	0.62	0.61
MAE	24.22	9.01	6.81	8.49	8.49	14.24	0.16	0.13	0.13	0.12	0.11	0.11	12.41	13.10	10.80	12.87	12.87	14.29
RMSE	29.91	11.89	10.06	11.97	11.97	18.18	0.20	0.18	0.17	0.16	0.15	0.15	17.19	18.90	17.59	18.54	18.54	18.83
RAE (%)	71.73	26.70	20.17	25.14	25.14	42.19	70.76	56.71	54.72	50.16	48.06	48.69	62.10	65.56	54.04	64.43	64.44	71.52
RRSE (%)	73.84	29.36	24.85	29.54	29.54	44.89	73.79	65.45	62.00	60.42	57.07	54.46	72.96	80.21	74.67	78.69	78.69	79.94
28 days, 5 test day records (TF)																		
r	0.73	0.96	0.96	0.96	0.96	0.94	0.73	0.81	0.82	0.81	0.81	0.81	0.66	0.62	0.67	0.61	0.61	0.58
MAE	23.84	8.32	8.72	8.41	8.41	12.49	0.16	0.13	0.12	0.12	0.12	0.13	12.74	12.80	10.23	12.80	12.80	13.83
RMSE	29.24	10.96	11.73	11.05	11.05	15.78	0.19	0.16	0.16	0.16	0.16	0.16	17.39	18.42	17.90	18.42	18.42	18.91
RAE (%)	69.34	24.18	25.35	24.45	24.45	36.31	65.87	53.06	49.31	50.66	50.80	53.85	65.17	65.49	52.32	65.50	65.50	70.73
RRSE (%)	71.57	26.82	28.71	27.04	27.04	38.62	69.09	58.74	57.08	57.63	57.64	58.79	74.57	79.01	76.74	78.98	78.98	81.08

* GP, Gaussian processes; LR, Linear regression; SMOreg, SMOregression; M5, M5Rules; M5P, M5P tree; RF, Random Forest. r, correlation coefficient; MAE, Mean Absolute Error; RMSE, Root Mean Squared Error; RAE, Relative Absolute Error; RRSE, Root Relative Squared Error.

Table 2. Comparison of actual values of Total Milk Yield, Peak Yield and Time to Peak Yield with machine learning algorithm estimates.

	7 days, 5 test day records (SF)			7 days, 20 test day records (ST)			28 days, 5 test day records (TF)		
	TMY (l)	PY (l)	TPY (d)	TMY (l)	PY (l)	TPY (d)	TMY (l)	PY (l)	TPY (d)
Actual	111.0ᵃ	0.97	37.6ᵃ	111.0	0.97ᵃ	37.6ᵃ	111.0	0.97ᵃ	37.6ᵃ
GP	87.9ᵇ	0.98	34.2ᵃ	121.0	1.0ᵃ	38.2ᵃ	120.0	4.78ᵇ	38.8ᵃ
LR	91.4ᵇ	0.92	33.5ᵃ	111.0	0.47ᵇ	37.9ᵃ	109.0	0.97ᵃ	37.7ᵃ
SMO	103.9ᵃ	0.88	24.0ᵇ	108.0	0.47ᵇ	24.7ᵇ	107.0.	0.95ᵃ	26.7ᵇ
M5	91.4ᵇ	0.92	34.1ᵃ	111.0	0.47ᵇ	38.5ᵃ	108.0	0.97ᵃ	37.7ᵃ
M5P	91.4ᵇ	0.92	34.1ᵃ	111.0	0.47ᵇ	36.6ᵃ	109.0	0.97ᵃ	37.7ᵃ
RF	92.1ᵇ	0.91	35.4ᵃ	111.0	0.47ᵇ	40.0ᵃ	108.0	0.96ᵃ	38.1ᵃ
S.E.M	3.86	0.03	1.85	4.01	0.02	2.35	3.93	0.03	2.22
P-value	0.003	0.05	0.001	0.38	0.001	0.004	0.26	0.001	0.008

*S.E.M. standard error of the mean; GP, Gaussian processes; LR, Linear regression; SMOreg, SMOregression; M5, M5Rules; M5P, M5P tree; RF, Random Forest.

those provided by the ST and TF schemes. This reinforces the idea that while simplified recording methods might reduce costs, the quality and quantity of the data are crucial for accurate predictions of key lactation characteristics, such as total milk production, peak production, and time to peak. From a practical standpoint, the TF method emerges as a highly efficient alternative due to its balance between cost-effectiveness and the accuracy of predictions. While more data-intensive methods like VRS provide marginally better accuracy, TF offers a more sustainable solution for commercial operations, where reducing operational costs is a priority.

These results are aligned with previous research that stresses the importance of both the data collection process and the algorithm selection when employing artificial intelligence in agricultural settings. The successful application of Machine Learning algorithms in this context opens doors for future studies focusing on further improving prediction models and exploring the integration of additional factors, such as environmental and management variables, to enhance the precision of lactation predictions.

5   Conclusions

The results of the current study demonstrate that ML algorithms are effective tools for managing sheep milk production by providing accurate estimates of lactation curve characteristics. Given the costs associated with measuring and recording milk production, the use of the TF system, which measures milk production each month, provides a cost-effective alternative that optimizes management practices by maximizing the information available while minimizing the operating costs of dairy sheep farms. Future research should focus on refining these algorithms for greater applicability to different sheep breeds and environmental conditions, with integration into practical farm management systems.

Acknowledgments. Guevara L. would like to thank the Coordenação de Aperfeiçoamento de Pessoal de Nível Superior - Brasil (CAPES) – Finance Code 001 for supporting the academic stay under the supervision of Dr. Juan Carlos Ángeles Hernández and the Conselho Nacional de Desenvolvimento Científico e Tecnológico - Brasil (CNPq) for funding the doctoral studies.

Disclosure of Interests.   The authors have no competing interests to declare that are relevant to the content of this article.

References

Alwadi, M., Alwadi, A., Chetty, G., Alnaimi, J.: Smart dairy farming for predicting milk production yield based on deep machine learning. Int. J. Inf. Technol. **16**(7), 4181–4190 (2024)

Anderson, S.M., Mao, I.L., Gill, J.L.: Effect of frequency and spacing of sampling on accuracy and precision of estimating total lactation milk yield and characteristics of the lactation curve1. J. Dairy Sci. **72**, 2387–2394 (1989). https://doi.org/10.3168/jds.S0022-0302(89)79371-1

Dallago, G.M., et al.: Predicting first test day milk yield of dairy heifers. Comput. Electron. Agric. **166**, 105032 (2019)

Guevara, L., et al.: Application of machine learning algorithms to describe the characteristics of dairy sheep lactation curves. Animals **13**(17), 2772 (2023)

Ince, D., Sofu, A.: Estimation of lactation milk yield of Awassi sheep with artificial neural network modeling. Small Rumin. Res. **113**(1), 15–19 (2013)

Mammadova, N., Keskin, İ: Application of the support vector machine to predict subclinical mastitis in dairy cattle. Sci. World J. **1**, 603897 (2013)

McGill, D., Thomson, P.C., Mulder, H.A., Lievaart, J.: Modification of lactation yield estimates for improved selection outcomes in developing dairy sectors: Conference of the Association for the Advancement of Animal Breeding and Genetics. Presented at the Conference of the Association for the Advancement of Animal Breeding and Genetics -AAABG 2013, AAABG, Napier, New Zealand, New Zealand, 20–23 October 2013, pp. 74–77 (2013)

McGill, D.M., Thomson, P.C., Mulder, H.A., Lievaart, J.J.: Strategic test-day recording regimes to estimate lactation yield in tropical dairy animals. Genet. Sel. Evol. **46**, 78 (2014). https://doi.org/10.1186/s12711-014-0078-0

Nguyen, Q.T., Fouchereau, R., Frénod, E., Gerard, C., Sincholle, V.: Comparison of forecast models of production of dairy cows combining animal and diet parameters. Comput. Electron. Agric. **170**, 105258 (2020)

Lenth, R.: Emmeans: estimated Marginal Means, aka Least-Squares Means. R package version 1.10.3, 2024. https://CRAN.R-project.org/package=emmeans

Sargent, F.D., Lytton, V.H., Wall, O.G., Jr.: Test interval method of calculating dairy herd improvement association records. J. Dairy Sci. **51**(1), 170–179 (1968)

Silvestre, A.M., Petim-Batista, F., Colaço, J.: The accuracy of seven mathematical functions in modeling dairy cattle lactation curves based on test-day records from varying sample schemes. J. Dairy Sci. **89**(5), 1813–1821 (2006)

Schaeffer, L.R., Jamrozik, J.: Multiple-trait prediction of lactation yields for dairy cows. J. Dairy Sci. **79**, 2044–2055 (1996). https://doi.org/10.3168/jds.S0022-0302(96)76578-5

Val-Arreola, D., Kebreab, E., Dijkstra, J., France, J.: Study of the lactation curve in dairy cattle on farms in central Mexico. J. Dairy Sci. **87**(11), 3789–3799 (2004)

Wasike, C.B., Kahi, A.K., Peters, K.J.: Modelling of lactation curves of dairy cows based on monthly test day milk yield records under inconsistent milk recording scenarios. Animal **5**, 1780–1790 (2011). https://doi.org/10.1017/S175173111100095

Author Index

© The Editor(s) (if applicable) and The Author(s), under exclusive license
to Springer Nature Switzerland AG 2025
L. Martínez-Villaseñor et al. (Eds.): MICAI 2024 Workshops, LNAI 15465, pp. 201–202, 2025.
https://doi.org/10.1007/978-3-031-83882-8